Cybersecurity

Artificial Intelligence (AI): Elementary to Advanced Practices

Series Editors: Vijender Kumar Solanki, Zhongyu (Joan) Lu, and Valentina E. Balas

In the emerging smart city technology and industries, the role of artificial intelligence is becoming increasingly prominent. This book series aims to cover the latest AI work, to help new users to solve existing problems and experienced AI practitioners pursue new avenues in the AI domain. The series covers recent work carried out in AI and its associated domains, including Logics, Pattern Recognition, NLP, Expert Systems, Machine Learning, Block-Chain, and Big Data. The range of work of AI is both extensive and profound. As a result, the series aims to cover the latest evolving AI trends, to provide fresh insights to those new to the field, current practitioners, students, and researchers.

Cyber Defense Mechanisms
Security, Privacy, and Challenges
Gautam Kumar, Dinesh Kumar Saini, and Nguyen Ha Huy Cuong

Artificial Intelligence Trends for Data Analytics Using Machine Learning and Deep Learning Approaches
K. Gayathri Devi, Mamata Rath, and Nguyen Thi Dieu Linh

Transforming Management Using Artificial Intelligence Techniques
Vikas Garg and Rashmi Agrawal

AI and Deep Learning in Biometric Security
Trends, Potential, and Challenges
Gaurav Jaswal, Vivek Kanhangad, and Raghavendra Ramachandra

Enabling Technologies for Next Generation Wireless Communications
Edited by Mohammed Usman, Mohd Wajid, and Mohd Dilshad Ansari

Artificial Intelligence (AI)
Recent Trends and Applications
Edited by S. Kanimozhi Suguna, M. Dhivya, and Sara Paiva

Deep Learning for Biomedical Applications
Edited by Utku Kose, Omer Deperlioglu, and D. Jude Hemanth

Cybersecurity
Ambient Technologies, IoT, and Industry 4.0 Implications
Gautam Kumar, Om Prakash Singh, and Hemraj Saini

For more information on this series, please visit: https://www.routledge.com/Artificial-Intelligence-AI-Elementary-to-Advanced-Practices/book-series/CRCAIEAP

Cybersecurity
Ambient Technologies, IoT, and Industry 4.0 Implications

Edited by

Gautam Kumar
Om Prakash Singh
Hemraj Saini

CRC Press is an imprint of the
Taylor & Francis Group, an **informa** business

First edition published 2021
by CRC Press
6000 Broken Sound Parkway NW, Suite 300, Boca Raton, FL 33487-2742

and by CRC Press
2 Park Square, Milton Park, Abi ngdon, Oxon, OX14 4RN

© 2022 Taylor & Francis Group, LLC

CRC Press is an imprint of Taylor & Francis Group, LLC

Reasonable efforts have been made to publish reliable data and information, but the author and publisher cannot assume responsibility for the validity of all materials or the consequences of their use. The authors and publishers have attempted to trace the copyright holders of all material reproduced in this publication and apologize to copyright holders if permission to publish in this form has not been obtained. If any copyright material has not been acknowledged please write and let us know so we may rectify in any future reprint.

Except as permitted under U.S. Copyright Law, no part of this book may be reprinted, reproduced, transmitted, or utilized in any form by any electronic, mechanical, or other means, now known or hereafter invented, including photocopying, microfilming, and recording, or in any information storage or retrieval system, without written permission from the publishers.

For permission to photocopy or use material electronically from this work, access www.copyright.com or contact the Copyright Clearance Center, Inc. (CCC), 222 Rosewood Drive, Danvers, MA 01923, 978-750-8400. For works that are not available on CCC please contact mpkbookspermissions@tandf.co.uk

Trademark notice: Product or corporate names may be trademarks or registered trademarks and are used only for identification and explanation without intent to infringe.

Library of Congress Cataloging-in-Publication Data
Names: Kumar, Gautam, 1990- editor. I Singh, Om Prakash, editor. I Saini,
Hemraj, 1977- editor.
Title: Cybersecurity : ambient technologies, IoT, and industry 4.0
implications / edited by Gautam Kumar, Om Prakash Singh, Hemraj Saini.
Other titles: Cybersecurity (CRC Press)
Description: First edition. I Boca Raton : CRC Press, 2022. I Series:
Artificial intelligence (AI) : elementary to advanced practices I
Includes bibliographical references and index.
Identifiers: LCCN 2021019345 (print) I LCCN 2021019346 (ebook) I ISBN
9780367702168 (hbk) I ISBN 9780367702175 (pbk) I ISBN 9781003145042 (ebk)
Subjects: LCSH: Internet of things--Security measures. I Industry
4.0--Security measures.
Classification: LCC TK5105.8857 .C96 2022 (print) I LCC TK5105.8857
(ebook) I DDC 005.8--dc23
LC record available at https://lccn.loc.gov/2021019345
LCebook record available at https://lccn.loc.gov/2021019346

ISBN: 978-0-367-70216-8 (hbk)
ISBN: 978-0-367-70217-5 (pbk)
ISBN: 978-1-003-14504-2 (ebk)

DOI: 10.1201/9781003145042

Typeset in Times
by SPi Technologies India Pvt Ltd (Straive)

The editors are thankful to the authors and reviewers of the concerned chapters who contributed to this book with their scientific work and useful comments, respectively.

Contents

Preface..ix

Editors...xiii

Contributors xv

Chapter 1 General and Specific Security Services, Risks, and Their Modeling 1

Suman De

Chapter 2 Vulnerability and Attack Detection Techniques: Intrusion Detection System 17

Dinesh Kumar Saini and Jabar H. Yousif

Chapter 3 Digital Rights Management in a Computing Environment 27

Maram Bani Younes and Nameer N. El-Emam

Chapter 4 Trade-Offs and Vulnerabilities in IoT and Secure Cloud Computing 43

Suman De

Chapter 5 Location and Availability Protections in Smart Mobility 65

Praveen Gupta and Reena Sharma

Chapter 6 Digital Forensics Cryptography with Smart Intelligence 83

Samaya Pillai, Venkatesh Iyengar, and Abhijit Chirputkar

Chapter 7 Transmission Modeling on Malware Attack through IoTs 103

Yerra Shankar Rao, Binayak Dihudi, and Tarini Charan Panda

Chapter 8 Rice Plant Disease Detection Using IoT 119

FarjanaYeasmin Trisha and Mahmudul Hasan

Chapter 9 Secure Protocols for Biomedical Smart Devices...........................131

Poonam Sharma, Prabhjot Kaur, and Kamaljit Singh Saini

Chapter 10 Access Control Mechanism in Health Care Information System..149

Bipin Kumar Rai and Tanu Solanki

viii Contents

Chapter 11 Privacy Preservation Tools and Techniques in Artificial
Intelligence 161
Raneem Qaddoura and Nameer N. El-Emam

Chapter 12 Web Security Vulnerabilities: Identification, Exploitation,
and Mitigation.. 183
*Sachin Kumar Sharma, Dr. Arjun Singh, Dr. Punit Gupta,
and Dr. Vijay Kumar Sharma*

Index... 219

Preface

Humankind has ventured into the time of the Industrial Revolution 4.0. Industry 4.0 could be a standard term to portray the fourth-generation Industrial Revolution that we are encountering these days. With each passing day, innovations such as cloud computing, the IoT, and mechanical technology are disturbing the conventional manufacturing process. With computerization, the IoT, and data analytics making production methods more and more intelligent, smart, and productive, the Industrial Revolution 4.0 has been evidently apparent. To all intents and purposes, this transitional move to digitization and robotization is known as "Industry 4.0."

However, this stage of the industrial revolution has introduced challenges, especially in terms of cybersecurity threats. Around the world, cybersecurity specialists have been concerned around the suggestions of Industry 4.0. In an age where everything is hyper-connected, industries have recently become more defenseless than ever. In other words, digitally connected industries are more vulnerable to assailants who are looking to abuse assets and information. As a result, the need for compelling cybersecurity measures inside an IIoT-enabled generation environment is posing a genuine threat to new-age industrial revolution systems and methods.

Nowadays, smart manufacturing plants and supply chains are consistently associated by means of an Industrial Internet of Things (IIoT) that makes use of IP addresses to associate and communicate inside and outside the production line. These internet-connected gadgets are always vulnerable to unauthorized access without legitimate cybersecurity measures.

Therefore, we attempt to provide a significant effort to identify and combat these risks, in the form of present book, *Cybersecurity: Ambient Technologies, IoT & Industry 4.0 Implications*. The book contains 12 chapters.

Chapter 1 covers the general security services and modeling aspects associated with any system. It also looks at the latest research done on such services and security models. The chapter explores the risks associated with a system and how, for a specific use-case of virtual reality implementation, we help the reader to see what risks can be mapped out and tackled. This chapter acts as a baseline for an individual to understand security services, the risks associated with a system, and the security models available, which can be used while designing a system.

Chapter 2 discuses the IDS approach, and methodology and different types of computer and network attacks. The latest trends, issues, and future research issues regarding intrusion detection systems are also presented. The chapter addresses the emerging research issues in designing reliable and accurate ID systems.

Chapter 3 defines the ethics and techno-ethics concepts, along with the main challenges and issues for ethics in technology. The chapter explains the main applications of cybersecurity, considering two main types: secure web applications and secure mobile applications. Finally, it explores the ethics for cybersecurity applications. This includes user privacy, freedom of speech, intellectual property rights, legal protections, and responsibility for crimes.

Chapter 4 presents the dependency on the technology that defines cloud computing and the IoT. Consumers face challenges from the risk of security flaws surrounding data privacy, loss, theft of intellectual property, insecure APIs, etc. These can become huge risks for any organization, as described in the chapter.

Chapter 5 discusses the software's Internet of Things, hardware, telecommunication network, and related technology issues required for these projects. Smart mobility applications and data are integral parts of the system, and therefore, problems related to data and their privacy may arise. The chapter deals with the security concerns of the public, legal issues, etc. Finally, the chapter discusses various levels of regulatory requirement and their roles.

Chapter 6 discusses various models of digital forensics, underlying principles, the concept of cryptography, and its impact on the domain of digital forensics. The chapter also addresses the impact of related technologies on human life, as Industry 4.0 has had a tremendous impact on these areas, enabling them, as well as enforcing the acceptance of such smart technologies in various aspects of our lives.

Chapter 7 proposes a mathematical model for malware attacks through the IoT, considering proper vaccination. The existence and uniqueness of the model have been proven. The infection-free and endemic equilibrium points have been found successfully. This model explains how the basic reproduction number determines the local and global stability of the system in IoT devices. It adapts the vaccination-based IoT devices in the network for protecting against malware attacks. The numerical simulations are critically analyzed and performed to validate the developed model.

Chapter 8 endeavors to build up a mechanized framework on rice plant disease detection using the IoT, which recognizes the presence of infection in the leaves and soil. There are various factors that determine pH, temperature, moisture, DHT 11, and TCS 3200 to detect the current position of the soil and plant. Those sensor values are sent, using an Ethernet shield, to be stored on an IoT server. Here, we have created a framework to detect the current condition of the plants' environment. We then analyze the current situation of the environment by comparing it with the healthy plant's actual values and show it on our constructed platform.

Chapter 9 focuses on the security issues of e-health care systems and provides the methodology to assure security. Various security protocols have been proposed, including an energy-efficient routing protocol, a secure protocol for user authentication and key agreement, a node-to-node authentication protocol, eliminating the man-in-middle attack, a lightweight anonymous authentication protocol for network security, and a trust key management protocol for biomedical smart devices. Finally, the chapter also discusses security protocols for biomedical smart devices.

Chapter 10 discusses different access control mechanisms associated with the health care system. In today's world, data has become the most important thing. Therefore, data privacy and security are crucial. In any information system, the satisfactory security of data, as well as access control by the owner of the data, are the primary requisites. Health care information systems have very crucial data on the patient. Therefore, the chapter concentrates on electronic health record information systems, which require the development of a strong mechanism to protect unauthorized access to the data.

Preface

xi

Chapter 11 identifies approaches combining several classification techniques with PSO for intrusion detection. A general presentation of the PSO algorithm is provided. PSO-based techniques for intrusion detection are introduced and detailed. The most common datasets and evaluation measures are discussed. Finally, a summary discussion about PSO-based intrusion detection techniques is provided, along with possible directions and insights.

Chapter 12 introduces the various types of vulnerabilities and their mitigation techniques. The biggest challenge for an organization is to prevent the web applications or portals from unauthorized activities, because web applications are available to all users 24/7, through the internet. Therefore, violating an implicit or explicit security policy is discussed, to counter the system's hardware or software vulnerabilities.

We hope that the works published in this book will serve the communities concerned with cybersecurity, the IoT, and Industry 4.0.

Editors

Dr. Gautam Kumar is currently Associate Professor at CMR Engineering College, Hyderabad, India. He received his PhD in Computer Science and Engineering from Jaypee University of Information Technology, Himachal Pradesh, India, in 2017. He received his M.Tech from Rajasthan Technical University, in 2012, and B.E. from Rajiv Gandhi Proudyogiki Vishwavidyalaya, Madhya Pradesh, in 2005. He has more than 15 years of academic experience. His research interests are in the fields of Cryptography, Information Security, Algorithms Design and Analysis. He has published more than 45 research journal articles and conference papers in Science Citation, Scopus, and Indexed Journals. He has also served as a president of Institute's Innovation Council, Ministry of Human Resource Development (MHRD), India and acted a Convenor/SPOC to the Smart-India Hackathon.

Dr. Om Prakash Singh is a Postdoc Research Associate in Medical Device Engineering at the University of Edinburgh, Edinburgh, Scotland, UK. He received a BA in Science, an MA in Physics and Biomedical Engineering, and a PhD in Biomedical Engineering, in 2005, 2007, 2009, and 2019, respectively. He has six years of teaching experience and four years of research experience. He has also filed 1 IP and obtained 2 copyrights for his research work. He has been awarded 2 gold awards and 1 merit award for his research. His research interests include the development of handy medical devices using optical-based sensors, by deploying signal-processing algorithms, and machine learning techniques for the automatic classification of cardiopulmonary conditions. To date, he has authored and coauthored around 19 researched and reviewed manuscripts, with an accumulated impact factor of 14.573.

Dr. Hemraj Saini is Associate Professor in the Department of Computer Science and Engineering, Jaypee University of Information Technology-(H.P), India. He received a PhD in Computer Science from Utkal University, Bhubaneswar (Orissa) in 2012, an M.Tech in IT from Panjabi University, Patiala, Panjab in 2005, and a B.E. in CSE from the National Institute of Technology, H.P., in 2000. His research interests include Network Security, Information Security,

Cybersecurity, the Internet of Things (IoT), Cloud Computing, Big Data, etc. He was awarded an Academic Excellence Award for Projects, a Merit cum Scholarship Award, and a National Scholarship. He has more than 20+ years of teaching and R&D experience at the national (Rajasthan, Orissa, H.P.) and international levels (Libya). He has published more than 160 research papers in journals and 45 conferences of international repute. He is in editorial member of reputed journals and conferences, with a great number of research collaborations. He has produced five PhD candidates, 4 of whom are under supervision, guided 13 M.Tech student projects, and guided 45 UG projects.

Contributors

Abhijit Chirputkar
Symbiosis Institute of Digital and
  Telecom Management
Symbiosis International (Deemed
  University)
Pune, India
director@sidtm.edu.in

Dr. Arjun Singh
Computer and Communication
  Engineering Department
Manipal University Jaipur, India
 vitarjun@gmail.com

Binayak Dihudi
Department of Mathematics
Konark Institute of Science and
  Technology
Jatni, Bhubaneswar, India
bdihudi@gmail.com

Bipin Kumar Rai
IT department
ABES Institute of Technology
Ghaziabad, UP, India
bipinkrai@gmail.com

Dinesh Kumar Saini
Computer and Communication
  Engineering Department
Manipal University Jaipur, India
dineshkumar.saini@jaipur.manipal.edu

Farjana Yeasmin Trisha
East West University
Dhaka, Bangladesh
farjana186@gmail.com

Jabar H Yousif
Computing and Information Technology
  Department
 Sohar University, Oman
jyusif@su.edu.om

Dr. Kamaljit Singh Saini
University Institute of Computing
Chandigarh University
Punjab, India
kamaljit.cse@cumail.in

Mahmudul Hasan
Jahangirnagar University
Dhaka, Bangladesh
mahmudul2843@gmail.com

Maram Bani Younes
Information Technology, Philadelphia
  University
Amman, Jordan
mbaniyounes@philadelphia.edu.jo

Nameer N. El-Emam
Information Technology
Philadelphia University
Amman, Jordan
nemam@philadelphia.edu.jo

Dr. Prabhjot Kaur
IT Department
MSIT, GGSIP University
New Delhi
prabhjot.kaur@msit.in

Poonam Sharma
University Institute of Computing
Chandigarh University
Punjab, India
poonam4sharma1987@gmail.com

Praveen Gupta
Department of Computer Engineering
Poornima Institute of Engineering and
  Technology
Jaipur, India
praveen2gupta@gmail.com

Dr. Punit Gupta
Computer and Communication
 Engineering Department
Manipal University Jaipur, India
punitg07@gmail.com

Raneem Qaddoura
Information Technology
Philadelphia University
Amman, Jordan
rqaddoura@philadelphia.edu.jo

Reena Sharma
Department of Computer Engineering
Poornima College of Engineering
Jaipur, India
shreena275@gmail.com

Samaya Pillai
Symbiosis Institute of Digital and
 Telecom Management
Symbiosis International (Deemed
 University)
Pune, India
samaya.pillai@sidtm.edu.in

Sachin Kumar Sharma
Manipal University Jaipur and
 Cybersecurity Analyst at Dr CBS
 Cyber Security Services LLP
Jaipur, India
sachin_43721@yahoo.com

Suman De
Development Specialist
SAP Labs India Pvt. Ltd.,
Bangalore, India
suman.de@sap.com

Tarini Charan Panda
Department of Mathematics
Ravenshaw University
Cuttack, India
tc_panda@yahoo.com

Tanu Solanki
IT Department
Galgotia College of Engineering and
 Technology
Greater Noida, UP, India
tanucsengg@gmail.com

Venkatesh Iyengar
Symbiosis Institute of International
 Businesses
Symbiosis International (Deemed
 University)
Pune, India
venkatesh.iyengar@siib.ac.in

Dr. Vijay Kumar Sharma
Computer and Communication
 Engineering Department
Manipal University Jaipur, India
Vijaymayankmudgal2008@
 gmail.com

Yerra Shankar Rao
Department of Mathematics
Gandhi Institute of Excellent
 Technocrats
Ghangapatana, Bhubaneswar,
 India
Email: sankar.math1@gmail.com

1 General and Specific Security Services, Risks, and Their Modeling

Suman De
SAP Labs India Pvt. Ltd., India

CONTENTS

1.1 Introduction 2
1.2 Literature Survey in Security Modeling 2
1.3 General Security Services 4
    1.3.1 Confidentiality 4
    1.3.2 Data Integrity 5
    1.3.3 Authenticity 5
    1.3.4 Authorization 5
    1.3.5 Non-Repudiation 5
    1.3.6 Support Services 6
    1.3.7 Combinatorial Service 6
    1.3.8 Key Management 6
1.4 Security Modeling 6
    1.4.1 Bell-LaPadula Model (GeeksforGeeks) 6
    1.4.2 Biba Model (GeeksforGeeks) 8
    1.4.3 Clarke–Wilson Security Model 9
    1.4.4 Graham-Denning Model 10
    1.4.5 Harrison-Ruzzo-Ullman Model 10
    1.4.6 Brewer-Nash Model 11
1.5 Risks 11
    1.5.1 Improper Platform Usage 11
    1.5.2 Insecure Data Storage 11
    1.5.3 Insecure Communication 12
    1.5.4 Insecure Authentication 12
    1.5.5 Insufficient Cryptography 12
    1.5.6 Insecure Authorization 13
    1.5.7 Client Code Quality 13
    1.5.8 Code Tampering 13
    1.5.9 Reverse Engineering 14
    1.5.10 Extraneous Functionality 14

DOI: 10.1201/9781003145042-1

2   Cybersecurity

1.6   Use Case: Virtual Reality 14
1.7   Summary 15
References 15

1.1   INTRODUCTION

- Framework: We need an escape from the arrangement important to see how the framework carries on when exposed to its proposed working conditions, just as unintended info or working conditions. A framework model may be founded on a norms report indicating conduct necessities, a plan detail, or a particular form or set of variants of source code.
- Threat: An escape from assailants' computational assets and framework access is essential. For instance, network assailants may approach network messages, yet not reach the host to impart with the organization. On the other hand, they may have an unbounded capacity, yet a deficient computational capacity to break cryptography. Operating system assailants may have the option to put pernicious code into a client cycle however unfit to adjust the OS portion.
- Properties: We should characterize the properties that we would like to keep aggressors from disregarding. For every conduct, for example, a succession of data sources, yields, and state transformations, we should decide if the ideal security properties hold or fall flat.

In this chapter, we look at security strategy and security models and how specific security services work. A security strategy is an archive that communicates clearly and succinctly what the security systems are to accomplish. It's an assertion of the security we anticipate that the framework should implement.

A security model is a detail of a security strategy:

- It depicts the elements represented by the arrangement,
- It expresses the principles that comprise the arrangement.

There are different sorts of security models:

- Models can capture arrangements for secrecy (Bell-LaPadula) or honesty (Biba, Clark-Wilson).
- A few models apply to conditions with static arrangements (Bell-LaPadula), while others consider unique changes of access rights (Chinese Wall).
- Security models can be casual (Clark-Wilson), semiformal, or formal (Bell-LaPadula, Harrison-Ruzzo-Ullman).

1.2   LITERATURE SURVEY IN SECURITY MODELING

There has been a significant amount of research been done in this space to control the balances of progressive system intrusion and how to better reduce them. In this section, we look at work done by researchers and the business aspects that they have considered. We start with the works of Jing Jin and Meihui Shen in 2012, titled, "Analysis of Security Models Based on Multilevel Security Policy". The researchers break down the advancement pattern of staggered security models. The center of the staggered security strategy is to isolate data into various security levels and to receive diverse assurance measures as per the security level. This approach is broadly utilized in the military and business fields. This chapter presents three data security models dependent on the staggered security strategy: the Bell-La Padula (BLP) model, the Biba model, and the Clark-Wilson model. It identifies and examines the attributes of the three models.

In "From Secure Business Process Modeling to Design-Level Security Verification" (2017), Qusai Ramadan, Mattia Salnitriy, Danel Struber, Jan Jurgens, and Paolo Giorgini discuss a set of modeling activities at an architectural level. Following and coordinating security prerequisites all through the improvement cycle is a critical test in designing security. In socio-specialized frameworks, security needs for the authoritative and specialized parts of a framework are currently managed independently, potentially giving rise to misguided judgments and blunders. In this work, the authors present a model-based security designing structure to support the framework plan on an authoritative and specialized level. The key thought is to permit the specialists included to indicate security needs in the dialects they know about: business investigators use BPMN for procedural framework portrayals; framework engineers use UML to plan and execute the framework design. Security needs are captured employing the language augmentations SecBPMN2 and UMLsec. They give a model change to connect the probable gaps between SecBPMN2 and UMLsec. Utilizing UMLsec strategies, different security properties of the subsequent design can be confirmed. For a situation study, they show how their system can be applied.

In 2015, Shireesha Katam, Pavol Zavarsky, and Francis Gichohi discussed the applicability of domain-based security risk modeling to SCADA systems. Domain-Based Security (DBSy) is a model-based methodology created by the Defense Evaluation and Research Agency for the UK Ministry of Defense to investigate data security chances in a business setting, to provide immediate planning for risks and the security controls expected to oversee them. The customary DBSy display parcels business measures and a basic IT framework into consistent areas of predefined classification levels, to authorize limitations on the sharing of data. While imperatives on data sharing concern the need for data privacy, mechanical control frameworks principally require and depend on the ideal and right data. Thus, this short paper investigates the materialness of the DBSy demonstrating to SCADA modern control framework conditions in which honesty and the accessibility of data are significant for the right activity of the situation, the security of human lives, and counteracting harm to the climate. The researchers' models appear to show that using privacy-centered inheritance style DBSy can be extended to consider and address the uprightness and accessibility needs of modern control frameworks.

In their 2018 paper, "A Quantitative Security Metric Model for Security Controls: Secure Virtual Machine Migration Protocol as the Target of Assessment," Tayyaba Zeb, Muhammad Yousaf, Humaira Afzal, and Muhammad Rafiq argue that quantitative security measurements are attractive for estimating the presentation of data security controls. Security measurements help to settle on utilitarian and business choices for improving the exhibition and cost of the security controls. Be that as it may, characterizing endeavor-level security measurements was recently recorded as a difficult issue, in the InfoSec Research Council's difficult issues list. Practically all the endeavors to characterize total security measurements for big business security have not been shown to be productive. Simultaneously, with the development of the security business, administrative bodies have constantly emphasized building up quantifiable security measurements. The researchers address this need and propose a general security metric model that determines three quantitative security measurements—the Attack Resiliency Measure (ARM), Performance Improvement Factor (PIF), and Cost/Benefit Measure (CBM)—for estimating the presence of the security controls. For the viability assessment of the proposed security measurements, we took the protected virtual machine (VM) movement convention as the objective of the evaluation. Virtualization advancements are quickly changing the registering scene. Contriving security measurements for a virtualized climate is significantly more challenging. The secure virtual machine movement is a developing are, and no standard convention is accessible explicitly for secure VM relocation. The 2018 paper took the protected virtual machine movement convention as the objective of evaluation and applied the proposed relative security metric model for estimating the ARM, PIF, and CBM of the safe VM relocation convention.

Building as a reference point, the world of cybersecurity has seen various upgrades. The aforementioned works identify just some of the recent research published by IEEE and form a small part of what we cover as General Security Services and Security Modeling techniques in this chapter.

1.3 GENERAL SECURITY SERVICES

This section sums up the essential security benefits that can be utilized to protect data, or as a supporting defensive component against attacks. The distribution portrays the accompanying essential security as secrecy, trustworthiness, confirmation, source verification, approval, and non-disavowal. A scope of cryptographic and non-cryptographic devices might be utilized to help these administrations. While a solitary cryptographic system could give more than one assistance, it can't offer a wide range of assistance. The potential administrations that can be related within an overall cybersecurity situation are as follows:

1.3.1 CONFIDENTIALITY

When forestalling the revelation of data to unapproved parties is required, the property of secrecy is also required. Cryptography is utilized to scramble data, to make it confused to everybody except the individuals who are approved to see it. The cryptographic calculation and method of activity should be planned and actualized so that an unapproved group will be not able to identify the keys that have been related to the encryption or infer the data without utilizing the right keys.

General and Specific Security Services

1.3.2 Data Integrity

Information trustworthiness confirms that information has not been adjusted in an unapproved way after it was made, communicated, or put away. This implies that there has been no inclusion, erasure, or replacement carried out with the information. Advanced marks or message confirmation codes are cryptographic instruments that can be utilized to identify both inadvertent changes that may happen as a result of equipment disappointment or transmission issues and intentional adjustments that may be performed by an enemy. While non-cryptographic instruments can be utilized to identify coincidental adjustments, they are not solid for distinguishing conscious changes.

1.3.3 Authenticity

Authenticity denotes the condition of truly speaking to the real world, that is, of being right (really speaking to that which is claimed to be spoken to), particularly with information starting at a conclusive, definitive source. Cryptography can give two sorts of verification administration:

- Respectability verification can be utilized to confirm that no adjustment has been made to the information.
- Source verification can be utilized to confirm the character of who made the data, for example, the client or the framework.

Computerized marks or message confirmation codes are utilized frequently to give validation administrations. Key-arrangement procedures may likewise be utilized to offer this support.

1.3.4 Authorization

An authorization needs an approval to play out a security capacity design for the same. This security administration is frequently supported by cryptographic help. Approval is commonly allowed after the effective execution of a source confirmation administration.

1.3.5 Non-Repudiation

In key administration, the term non-disavowal alludes to the authority of an authentication subject using computerized signature keys and advanced endorsements to a public key. At the point when non-disavowal is needed for an advanced mark key, it implies that the mark that has been made by that key has the help of both the trustworthiness and source confirmation administrations of a computerized signature. The computerized mark may likewise demonstrate a dedication by the method of the authentication subject in a similar way that an archive with a transcribed mark would. Nonetheless, there are numerous angles to be considered in settling on a legitimate choice for non-disavowal, and this cryptographic component is viewed as just a single component to be utilized in that choice.

6 Cybersecurity

1.3.6 SUPPORT SERVICES

Supporting administrations are frequently needed for the fundamental cryptographic security administrations mentioned earlier. For instance, cryptographic help will regularly require administrations for key foundation and irregular number age, just as for the security of the cryptographic keys themselves.

1.3.7 COMBINATORIAL SERVICE

A mix of the more than six security administrations is emphatically exhorted. When planning a safe framework, architects for the most part start by figuring out which security frameworks are needed to ensure the data that will be contained and prepared by the framework. When the administrations have been resolved, the components that will best offer these types of assistance are thought of. A portion of the instruments picked probably won't be cryptographic. For instance, actual safety efforts, for example, distinguishing proof identifications, or biometric ID gadgets might be utilized to restrict admittance to certain information for privacy purposes. Nonetheless, cryptographic instruments that incorporate calculations, keys, or other key material are commonly the most practical strategies for keeping data secure.

1.3.8 KEY MANAGEMENT

The right administration of cryptographic keys is fundamental to the degree of security that may be accomplished in a framework through cryptography. This feasible security relies upon different factors, for example, the engineering of the cryptographic framework or the applied blend of components and their characteristic strength against attacks.

So, what is the connection between the security level of a framework, cryptographic keys, and cryptographic systems? All of the scrambled data in a framework is ensured by cryptographic keys. This protection stays operational as long as the cryptographic keys have not been undermined. At the end of the day, to ensure the essential security administrations provided by cryptography, we have to do everything needed to guarantee that the defensive instruments for dealing with the keys safely don't fizzle.

1.4 SECURITY MODELING

The foregoing services indicate the possibility of using the parameters for security services and design or model systems to better handle external breaches. The security models are taken into consideration to ensure the objectives of keeping a system secure. We look at the most important traditional security models in this section, before exploring other security models.

1.4.1 BELL-LAPADULA MODEL (GEEKSFORGEEKS)

Developed by scientists David Elliot Bell and Leonard J. LaPadula, this model is known as the Bell-LaPadula Model (see Figure 1.1). It is utilized to maintain the

General and Specific Security Services

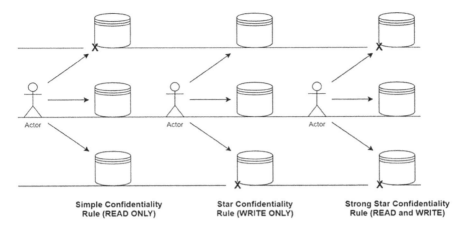

FIGURE 1.1 Bell-LaPadula Model.

(Source: GeeksforGeeks, Introduction to Classic Security Models)

confidentiality of security. Here, the arrangement of subjects (users) and objects (files) is coordinated in a non-optional style, concerning various layers of secrecy. Some important points about this model are as follows:

- It is the main numerical model with a staggered security strategy that is utilized to characterize the idea of a safe state machine and models of access and plot rules of access.
- It is a state m/c model that authorizes the classification parts of the entrance model.
- The model spotlights guaranteeing that subjects with various clearances (top secrecy, secrecy, private) are appropriately verified by having the vital exceptional status, need to know, and formal access endorsement before getting to an item that is under various grouping levels (top secrecy, secrecy, secret).
- The principles of the Bell-LaPadula Model:
  - Basic Confidentiality Principle: The Simple Confidentiality Rule states that the subject can read only the records on the same layer of secrecy and the lower layer of secrecy, however, not the upper layer of secrecy. Because of this, we call this standard NO READ-UP.
  - Star Confidentiality Principle: The Star Confidentiality Rule states that the subject can write only the records on the same layer of secrecy and the upper layer of secrecy, but not the lower layer of secrecy. Because of this, we call this standard NO WRITE-DOWN.
  - Strong Star Confidentiality Principle: The Strong Star Confidentiality Rule is deeply verified and grounded. It states that the subject can read and write the records only on the same layer of secrecy and not the upper layer of secrecy or the lower layer of secrecy. Because of this, we call this standard NO READ WRITE UP DOWN.

- Solid star property rule: This rule states that a subject with peruse and compose abilities can play out those capacities only at a similar security level, not much higher, and nothing lower.
- Peacefulness rule: Subjects and items cannot change their security levels once they have been started up (made).
- All MAC frameworks depend on the Bell-LaPadula model, due to its staggered security levels.
- Generally embraced by government organizations, this model is in planning for use by the US government.

1.4.2 Biba Model (GeeksforGeeks)

Developed by scientist Kenneth J. Biba, this model is called the Biba Model (see Figure 1.2). It is utilized to maintain the integrity of security. Here, the order of subjects (users) and objects (files) are coordinated in a non-optional design, concerning various layers of secrecy. This works the conversely to the Bell-LaPadula Model. The basic characteristics of the Biba Integrity Model are:

- It is created after the Bell-LaPadula Model.
- It tends to the trustworthiness of information, in contrast to Bell-Lapadula, which tends to secrecy.
- It utilizes a cross-section of uprightness levels, in contrast to Bell-Lapadula, which utilizes a grid of security levels.
- It is a data stream model, like Bell-Lapadula, because it is generally concerned with information moving, starting with one level then onto the next.
- The principles of the Biba Model:
  - Basic Integrity Principle: The Simple Integrity Rule states that the subject can only read the files on the same layer of secrecy and the upper layer of

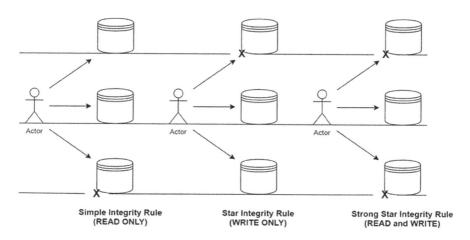

FIGURE 1.2 Biba integrity Model.

(Source: GeeksforGeeks, Introduction to Classic Security Models)

secrecy, but not the lower layer of secrecy. Due to this, we call the rule NO READ DOWN.
- Star Integrity Principle: The Star Integrity Rule states that the subject can only write the files on the same layer of secrecy and the lower layer of secrecy, but not the upper layer of secrecy. As a result, we call this rule NO WRITE-UP.
- Strong Star Integrity Rule: This is the same as the Star Integrity Rule, but in addition, it holds read and write capabilities and can only perform both the actions at the same security level in reference to subject approval and object classifications.

1.4.3 Clarke–Wilson Security Model

The Clarke-Wilson Model is a highly secure model, developed after Biba (see Figure 1.3). It focuses on addressing the integrity aspects of the knowledge available with a system. Other significant characteristics of this model are:

- The isolation of information into one subject that should be exceptionally ensured, alluded to as an obliged information item (CDI), and another subset that doesn't need a significant level of security, alluded to as unconstrained information items (UDI).
- Segments:
    - Subjects (clients): are dynamic specialists.
    - Transformation Procedures (TPs): the s/w systems, for example, perused, compose, that play out the necessary procedure for the subject (client).

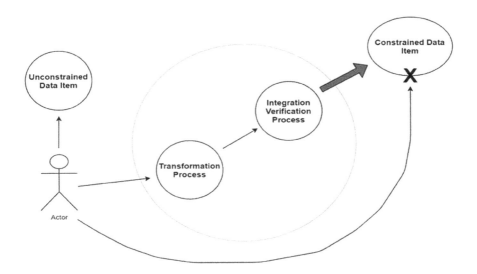

FIGURE 1.3 Clarke-Wilson Security Model.

(Source: GeeksforGeeks, Introduction to Classic Security Models)

10   Cybersecurity

- ○ Compelled Data Items (CDI): information that can be adjusted simply by TPs.
- ○ Unconstrained Information Things (UDI): information that can be controlled by subjects through crude read/compose tasks.
- ○ Integrity Verification Procedure (IVP): programs that run occasionally to check the consistency of CDIs with outer reality. These respectability rules are generally characterized by vendors.
- Integrity objectives of the Clark-Wilson Model
  - ○ Keep unapproved clients from making changes (tended to by the Biba Model).
  - ○ Partition obligations, to keep approved clients from making ill-advised changes.
  - ○ All-around shaped exchanges: maintain inward and outward consistency. For example, it is a progression of tasks that are completed to move the information from one reliable state to the next.

A security model guides the theoretical objectives of the approach to data framework terms by indicating unequivocal information structures and methods that are important to authorize the security strategy. In general, one can consider a security model as a framework designed scientifically and logically, and afterward created by developers through executing code. Along with the foregoing three models, others are used while modeling such systems:

1.4.4   GRAHAM-DENNING MODEL

This model characterizes a group of fundamental rights, as far as orders that a particular subject can execute in an article. It proposes the eight basic assurance rights or rules of how these sorts of functionalities should occur safely.

- Step-by-step instructions to safely make an article.
- Step-by-step instructions to safely make a subject.
- The most effective method to safely erase an item.
- Instructions to safely erase a subject.
- Instructions to give read admittance rights.
- Instructions to give award access rights.
- Instructions to give erase access rights.
- Instructions to give move access rights.

1.4.5   HARRISON-RUZZO-ULLMAN MODEL

The HRU security model (Harrison, Ruzzo, Ullman Model) is a working system-level PC security model that deals with access rights. The framework is based around the possibility of a limited arrangement of methods being accessible to alter the entrance privileges of a subject (s) on an object (o). The model additionally addresses the potential outcomes and constraints of demonstrating the well-being of a framework utilizing a calculation (Wikipedia).

General and Specific Security Services

1.4.6 BREWER-NASH MODEL

This security model is also known as the Chinese Wall security model. It provides access controls that can change powerfully, depending on a client's past activities. The fundamental objective of this model is to secure against irreconcilable situations by the client's entrance endeavors. It depends on the data stream model, where no data can stream among subjects and articles in a manner that would bring about an irreconcilable circumstance. The model expresses that a subject can keep in touch with an article if, and only if, the subject cannot peruse another item that is in an alternate informational index.

Security architects also look at the techniques of Covert Channels, Information Flow Models, the Access Control Matrix, and so on for security modeling purposes; these techniques are not covered in detail as part of the scope of this chapter.

1.5 RISKS

With all applications, there are numerous related risks. According to the Open Web Application Security Project (OWASP), we have Improper Platform Usage, Insecure Data Storage, Insecure Communication, Insecure Authentication, Insufficient Cryptography, Insecure Authorization, Client Code Quality, Code Tampering, Reverse Engineering, and Extraneous Functionality. We take a look at those dangers in this section.

1.5.1 IMPROPER PLATFORM USAGE

Infringement of distributed rules: All stages have created rules for this. If an application negates the accepted procedures suggested by the maker, it will present this danger.

Infringement of show or normal practice: Not all prescribed procedures are arranged in producer direction. In certain cases, there are accepted prescribed procedures that are normal in versatile applications.

Unexpected Misuse: Some applications expect to make the best decision, yet they get some piece of the execution wrong. This could be a basic bug, such as setting an unacceptable banner on an API call, or it very well may be a misconception of how the assurances work.

1.5.2 INSECURE DATA STORAGE

Here, an attacker genuinely breaches a cell phone. The enemy connects the virtual reality (VR) gadget to a PC, with uninhibitedly accessible programming. These apparatuses permit the enemy to see all outsider application indexes that frequently contain recognizable data or other sensitive data resources. An enemy may develop malware or adjust a genuine application to take such data resources. Uncertain information stockpiling weaknesses normally lead to accompanying industry hazards for the company that uses the dangerous application:

- Identity Theft
- Scam

12   Cybersecurity

- Image Violation
- External Policy Violation (PCI)
- Collateral Damage

1.5.3   INSECURE COMMUNICATION

When planning a versatile application, information is regularly traded in a customer worker design. At the point when the arrangement sends its information, it must cross the cell phone's transporter organization and the web. Attackers may abuse weaknesses to block sensitive information while it's traversing the wire. At any rate, the attempt to block sensitive information through a correspondence direct will bring about a security infringement.

The infringement of a client's secrecy may result in:

- Identity Theft
- Scam
- Image Violation

1.5.4   INSECURE AUTHENTICATION

Threat specialists who misuse confirmation weaknesses commonly do so through computerized attacks that utilize accessible or exceptionally assembled devices. When the attacker sees how the verification process is powerless, they simulate or sidestep validation by submitting administration solicitations to the VR gadget's back-end worker and sidestep any immediate communication with the VR gadget. The business effect of false confirmation commonly brings about:

- Image Violation
- Information Theft
- Unauthorized Access to Data

1.5.5   INSUFFICIENT CRYPTOGRAPHY

Threat specialists incorporate the accompanying: anybody with actual admittance to information that has been encoded inappropriately, or portable malware following up for an enemy's sake. Attack vectors incorporate the accompanying: unscrambling of information through actual admittance to the gadget or organization traffic catch, or malevolent applications on the gadget with admittance to the scrambled information. This weakness can have distinctive business impacts. Regularly, broken cryptography will bring about:

- Privacy Violations
- Information Theft
- Code Theft
- Intellectual Property Theft
- Image Violation

General and Specific Security Services

1.5.6 INSECURE AUTHORIZATION

Threat specialists who misuse approval weaknesses normally do so through robotized attacks that utilize accessible or uniquely assembled devices. When the attacker sees that the approval conspires is helpless, they sign in to the application as a real client. They effectively pass confirmation control. Once past confirmation, they commonly power peruse to a weak endpoint to execute authoritative usefulness. If a client (secret or checked) can execute overfavored usefulness, the business may encounter:

- Image Violation
- Scam
- Information Theft

1.5.7 CLIENT CODE QUALITY

Threat specialists incorporate elements that can pass untrusted contributions to strategy calls made inside versatile code. These sorts of issues are not security issues all by themselves; however, they lead to security weaknesses. An assailant will regularly misuse weaknesses in this class by providing painstakingly created contributions to the person in question. These data sources are sent to code that lives inside the cell phone where the misuse happens. Regular sorts of attacks misuse memory holes and cushion floods. The impact on a business from this classification of weaknesses fluctuates enormously, contingent on the idea of the endeavor. Helpless code quality issues that bring about far-off code execution could prompt accompanying business impacts:

- Information Theft
- Secrecy Violation
- Intellectual Property Theft

Different, less-serious specialized issues that fall into this classification may prompt corruption in execution, memory use, or helpless front-end engineering.

1.5.8 CODE TAMPERING

Normally, an attacker will abuse code change utilizing vindictive types of the applications facilitated in third-gathering application stores. The attacker may likewise fool the client into introducing the application through phishing attacks. Regularly, an attacker does the following to misuse this class:

- Make direct paired changes to the application bundle's center binary
- Make direct parallel changes to the assets inside the application's bundle
- Redirect or supplant framework APIs to catch and execute unfamiliar code that is malevolent

The business effect of code change normally brings about:

- Revenue loss because of theft
- Image violation

1.5.9 Reverse Engineering

The objective of the hacking network with Playstation VR is certainly not a secrecy: get the Sony gadget to chip away at different gadgets than the PS4. One evident advantage is that Playstation VR is perhaps the least expensive choice for VR today (clarified further in the use case, Section 1.7). Making it viable with PCs would give a pleasant section highlight VR for PC proprietors who would prefer not to go with the more costly choices, for example, HTC Vive or the Oculus Rift.

1.5.10 Extraneous Functionality

Normally, an intruder looks to comprehend superfluous usefulness inside a versatile application, to find concealed usefulness in back-end frameworks. The attacker ordinarily misuses superfluous usefulness straightforwardly from their frameworks with no contribution by end-clients. The attacker downloads and analyze the portable application inside its nearby climate. They look at log documents, design records, and maybe the paired application itself to find any shrouded switches or test code that was abandoned by the engineers. They abuse these switches and shrouded usefulness in the back-end framework to play out an attack. The business effect of superfluous usefulness incorporates:

- Unauthorized Access to Sensitive Functionality
- Reputational Damage
- Intellectual Property Theft

1.6 USE CASE: VIRTUAL REALITY

With different players coming into the market and enormous amounts of speculation, there has been incredible exploration done on the execution systems for VR. As an illustration, Oculus Rift has attracted consideration in recent years. Solidarity, CAD, and other systems have developed additionally after some time and added to the upgrading of the client experience of getting a reasonable perspective on environmental factors through VR.

Solidarity has presented underlying help for certain VR gadgets. This guide zeroes in on the Oculus group of VR gadgets, especially the Oculus Rift Development Kit 2 (DK2) and the shopper version of the Gear VR (a portable headset that requires a Samsung Galaxy S6, S6 Edge, S6 Edge+, or Note 5 handset). We are not zeroing in on the Note 4, which was recently upheld by the principal Innovator Edition of the GearVR. However, we expect that the VR tests will run on this gadget (though with lower execution) for those of you who have it.

Other VR head-mounted displays (HMDs) will likewise work with Unity, for example, the HTC Vive, and this documentation will be refreshed later on to cover extra VR stages. Such devices include a lot of possibilities for how breaches can happen. Every such implementation involves risks, and the aforementioned modeling aspects relate to how and where there is a possibility for a system to be compromised. As a follow-up to this chapter, the reader might take it as an exercise to figure out what risks are possible for a VR use case.

1.7 SUMMARY

This chapter covers a host of topics, providing the gist of the available services and aspects to be considered for security modeling. The topics are:

- Cybersecurity and its importance for the current scope of digitization and the industry-based implementation of various software products.
- The latest research done on cybersecurity and how these studies have created an impact as part of their contributions.
- The basic security services that are expected for a system to be considered well designed.
- The security services of Confidentiality, Integrity, Authenticity, Authorization, Keys Management, Support Services, and Non-Repudiation.
- Security models, especially the Bell-LaPadula, Biba Integrity, and Clarke-Wilson models.
- Risks that can affect a system, with a brief on each of them.
- The use-case of implementing VR, which is suggested as a task for the reader, an opening to explore associated risks for such a use case.

REFERENCES

De, S. (2015). "Addressing General Data Protection Regulations for Enterprise Applications Using Blockchain Technology." *International Journal of Applied Research on Information Technology and Computing*, vol. 10, no. 2, pp. 67–75. ISSN: 0975-8070. DOI: 10.5958/0975-8089.2019.00009.5

De, S., & Vijayakumaran, V. (2019,) "A Brief Study on Enhancing Quality of Enterprise Applications using Design Thinking." *International Journal of Education and Management Engineering*, vol. 9, no. 5, pp. 26–38. DOI: 10.5815/ijeme.2019.05.04

GeeksforGeeks. (2020, July 9). "Introduction to Classic Security Models." Available at: https://www.geeksforgeeks.org/introduction-to-classic-security-models/

Harris, Shon. (n.d.). "CISSP Certification All-in-One Exam Guide." Security Models and Architecture. Available at: https://media.techtarget.com/searchSecurity/downloads/29667C05.pdf

Jin, J., & Shen, M. (2012). *"Analysis of Security Models Based on Multilevel Security Policy."* *2012 International Conference on Management of e-Commerce and e-Government*, pp. 95–97. Beijing: IEEE. DOI: 10.1109/ICMeCG.2012.72

Katam, S., Zavarsky, P., & Gichohi, F. (2015). *"Applicability of Domain-Based Security Risk Modeling to SCADA Systems."* *2015 World Congress on Industrial Control Systems Security (WCICSS)*, pp. 66–69. London: Elsevier. DOI: 10.1109/WCICSS.2015.7420327

Kumar, G., Saini, D., & Cuong, N. (Eds.). (2021). *Cyber Defense Mechanisms*. Boca Raton: CRC Press. DOI: 10.1201/9780367816438

Mitchell, J., & Bau, J. (2011). "Security Modeling and Analysis." *IEEE Security & Privacy*, vol. 9, no. 03, pp. 18–25. DOI: 10.1109/MSP.2011.2

Pal, S. K., & De, S. (2015). "An Encryption Technique Based upon Encoded Multiplier with Controlled Generation of Random Numbers." *IJCNIS*, vol.7, no.10, pp.50-57. DOI: 10.5815/ijcnis.2015.10.06

Panjwani, M., & De, S. (2020). *"Study of Cloud Security in Hyper-Scalers."* *7th International Conference on Computing for Sustainable Global Development (INDIACom)*, pp. 29–34. New Delhi: IEEE. DOI: 10.23919/INDIACom49435.2020.9083727

Ramadan, Q., Salnitriy, M., Strüber, D., Jürjens, J., & Giorgini, P. (2017). "*From Secure Business Process Modeling to Design-Level Security Verification.*" *2017 ACM/IEEE 20th International Conference on Model Driven Engineering Languages and Systems (MODELS)*, pp. 123–133. Austin, TX: IEEE. DOI: 10.1109/MODELS.2017.10

The Open Group. (1997). "Authentication and Security Services, DCE 1.1." Available at: https://pubs.opengroup.org/onlinepubs/9668899/chap1.htm

Turner, D. M. (2017). "Applying Cryptographic Security Services – A NIST Summary." Available at: https://www.cryptomathic.com/news-events/blog/applying-cryptographic-security-services-a-nist-summary

Wikibooks. (n.d.). "Security Architecture and Design." Available at: https://en.wikibooks.org/wiki/Security_Architecture_and_Design/Security_Models

Wikipedia. (n.d.). "HRU Security." Available at: https://en.wikipedia.org/wiki/HRU_(security)

Zeb, T., Yousaf, M., Afzal, H., & Mufti, M. R. (2018, Aug.). "A Quantitative Security Metric Model for Security Controls: Secure Virtual Machine Migration Protocol as Target of Assessment." *China Communications*, vol. 15, no. 8, pp. 126–140. DOI: 10.1109/CC.2018.8438279

2 Vulnerability and Attack Detection Techniques
Intrusion Detection System

Dinesh Kumar Saini
Manipal University Jaipur, India

Jabar H. Yousif
Sohar University, Oman

CONTENTS

2.1	Introduction	18
2.2	Cybersecurity Services	20
2.3	Intrusion Detection System (IDS) Software Architecture	20
	2.3.1 IDS	20
	2.3.2 Detection Approach	21
	2.3.2.1 Anomaly Detection Approaches	21
	2.3.2.2 Misuse Detection Approaches	21
	2.3.3 Methodology Approach	21
	2.3.3.1 Signature-Base Detection (SBD)	21
	2.3.3.2 Anomaly-Based Detection (ABD)	21
	2.3.3.3 Stateful Protocol Analysis (SPA)	22
2.4	Classification of Computer and Network Attacks	22
	2.4.1 Attack Type	22
	2.4.1.1 DoS Attack	22
	2.4.1.2 Probing Attacks	23
	2.4.1.3 Number of Network Connections Required by the Attacks	23
2.5	Source of the Attack	23
	2.5.1 Environment	24
	2.5.1.1 Intrusion on Local Host Machine	24
	2.5.1.2 Network Intrusion	24
	2.5.1.3 Intrusion in Wireless Network	24
2.6	Automated Level	24
	2.6.1 Automation Attacks	24
	2.6.2 Semiautomated	24
	2.6.3 Manual Attacks	24

DOI: 10.1201/9781003145042-2

18   Cybersecurity

2.7   Intrusion Detection System Component 25
        2.7.1   Data Gathering Component 25
        2.7.2   Knowledge-Based Component 25
        2.7.3   Configuration Component..25
        2.7.4   Response Component..25
2.8   Conclusion..25
2.9   Future Work..26
References...26

2.1   INTRODUCTION

With the significant increase in the use of the Internet, social media, and data trans-mission, cybersecurity has become the primary concern to prevent cyberthreats and attacks, which are also increasing. Attackers use sophisticated techniques to target systems with vulnerabilities, whether those of large or small organizations or indi-viduals [1]. It is now necessary to focus on the importance of cybersecurity and tak-ing all possible steps to deal with and cyberthreats. Computer systems and networks such as the Internet have a broad application in different business areas, and it is rapidly increasing, leading to massive needs for security, integrity, privacy, and con-fidentiality of the data while the data is transmitted over the Internet to others. The traditional security available to protect computer systems and network systems, such as authentication devices, firewalls, and virtual private networks (VPNs), has many different types of vulnerability [2]. It cannot monitor advanced threats and new kinds of intrusion.

Cybersecurity is defined as a set of steps and technologies that help protect sensi-tive data, computer systems, networks, and software applications from cyberattacks. Attackers target a range of resources, including stored vulnerable systems and sensi-tive data, exploit resources, gain unauthorized access to systems and information, encode data encryption to extort money from victims, and disrupt systems' normal functioning and their operations. Attacks today have become more creative and sophisticated and can disrupt security programs and infiltrate systems. Therefore, it is challenging for security systems and analysts to track and resist attacks [3]. Figure 2.1 shows the cybersecurity market size of the project services in the United Kingdom from 2010 to 2017.

For this reason, security researchers have made efforts to improve security in computer systems and networks [4]. However, this issue still exists because of new daily vulnerabilities in system hardware and software, plus vulnerability in the updated release. To be more precise, no network is 100% secure, due to different types of new attacks. These attacks can be improved and their methods changed, when there is a unique signature of abnormal behavior in the database. Therefore, researchers have proposed different tools to help reduce an attack's effect to an acceptable level.

Researchers have divided the security operation life cycle into five phases: prepa-ration, dry run, execution, evaluation, and repetition [5]. The Intrusion Detection System (IDS) is a proposed solution available in a software or hardware base. IDS can monitor both the computer system and network traffic, using different detection

Vulnerability and Attack Detection Techniques 19

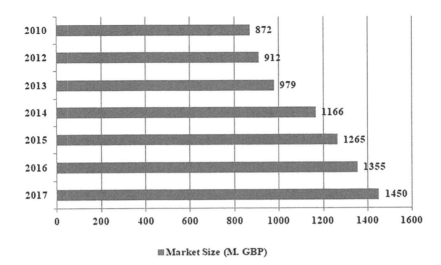

FIGURE 2.1 Example, the cybersecurity project services market size in the United Kingdom.

Source (https://www.statista.com/statistics/290005/uk-cyber-security-project-services-and-outsourcing-segment-size).

methods and approaches, and then report on it to the network administrator for decision-making.

Cintuglu et al. [6] examine the testbeds that apply to cyber-physical systems, such as industrial control systems and the power grid. They also investigate testbeds that utilize the target objectives, such as vulnerability analysis, education, and defensive mechanisms. Therefore, it is vital to improving the specific architectural decisions to align with particular target vulnerabilities.

Shone et al. [7] address the implementation of deep learning methods for network intrusion detection systems (NIDS), which defend computer networks. They review the issues regarding the feasibility and sustainability of current systems when responding to network requirements. They deploy a nonsymmetric deep autoencoder (NDAE) unsupervised method for feature learning and model classification. The proposed classifier is evaluated and benchmarked using KDD Cup '99 and NSL-KDD datasets. The results demonstrate a promising improvement for use in modern NIDSs, in comparison with current approaches.

M. Ferrag, M. et al. [8] survey the using of deep learning methods for cybersecurity intrusion detection and examined the datasets used in several studies. They compare 35 well-known cyber datasets and classify them into seven classes, including network traffic, electrical network, internet traffic, internet-connected devices, IoT traffic, virtual private network, and android apps. They provide a comparative study of seven deep learning approaches to examine the performance and the accuracy of each system under two different datasets: real traffic and the Bot-IoT.

Ahmad E. et al. [9] conduct a critical literature review to examine the current state of the research efforts in the field of cyber-physical attacks and to highlight future research directions and those areas needing more attention. The results show that the most extensive published research papers were focused on attack detection techniques (28%) and presenting cyber-physical attack case studies (20%). Also, the results show that 15% of the research is offered with work related to classification and modeling approaches.

2.2 CYBERSECURITY SERVICES

The increase in the number of attacks and threats has demonstrated the need for cybersecurity services in the public and private sectors. Therefore, the global cybersecurity services market is expanding rapidly, driven by various trends. Cybersecurity services help identify the most valuable assets, identify and manage risks, achieve compliance, and improve business performance. Current service companies focus on helping institutions, processes, technology, and individuals ensure their control and protect vital functions and software from external threats.

So, the most significant costs associated with entering the cybersecurity market are gaining experience through basic training and the development of solutions, including brochure production and product marketing. Cybersecurity technology is expanding rapidly, due to the increase in the Internet of Things (IoT) through the adoption of cloud computing services (Software as a Service, SaaS) over traditional services.

The following are examples of services that distributors provide reliable solutions and products:

- Network security monitoring and protection tools
- Firewall protection
- Data classification and encryption tools
- Antivirus and malware software
- Web vulnerability scanning tools
- Managed detection services
- Network defense wireless tools
- Packet sniffers
- Public key infrastructure services
- Penetration testing

2.3 INTRUSION DETECTION SYSTEM (IDS) SOFTWARE ARCHITECTURE

2.3.1 IDS

IDS can monitor all traffic in the network system and analyze it using different technical methodologies and approaches, to detect attacks or any traffic intruding on the network and report it to the network administrator to take appropriate action to block or allow the traffic. IDS only can monitor and

Vulnerability and Attack Detection Techniques

detect abnormal traffic. To block and prevent attacks, you need to use an intrusion prevention system (IPS).

2.3.2 DETECTION APPROACH

Most security researchers classify IDS approaches in two essentials categories:

2.3.2.1 Anomaly Detection Approaches

Anomaly detection approaches depended on analyzing the host, network connection, and user profiles to classify them either as normal or abnormal traffic. The main advantage of anomaly detection approaches is the ability to stop unexpected attacks. This approach's greatest drawback is a high rate of false alarms, which regard some traffic as actual attacks. Still, the traffic is a new or unusual use of normal legitimate network behaviors.

2.3.2.2 Misuse Detection Approaches

Misuse detection approaches work by using a database that stores exhaustive information about different types of system vulnerability and network attacks, defined by expert people in the network security field.

2.3.3 METHODOLOGY APPROACH

IDS normally uses three methods to detect intrusions. We classify these methods as follows:

2.3.3.1 Signature-Base Detection (SBD)

Signature-Base Detection is also called misuse detection or knowledge base detection. It works as an antivirus technique that detects the intrusion's pattern or signature and compares it with the library database to decide if this traffic is legitimate or not and report it to the network administrator. Using this technique, it is very easy to detect known attacks, because information about the attack is available in the database in contracts. It's challenging to detect unknown attacks, however, because the database stele is not yet updated [1].

2.3.3.2 Anomaly-Based Detection (ABD)

This method is also called behavior-based detection. It focuses on detecting both network traffic and computer intrusion or misuse by determining them to be normal or abnormal, using predefined rules created by the network administrator instead of using patterns or signatures. In these methods, the system can be used in several ways to detect positive traffic and report unwanted traffic to the network administrator [3].

TABLE 2.1
The advantages and disadvantages of IDS methodology

Methodology	Advantage	Disadvantage
Signature-based detection (SBD)	- Very simple method. - Powerful in detecting known attacks.	- Unable to detect unknown attacks. - Very difficult to keep signature pattern up to date. - Time-consuming to search in the database.
Anomaly-based detection (ABD)	- More powerful to detect new attacks and unexpected vulnerability. - Compatible with all operating system. - Easy to detect misuse of privileges.	- Due to constant changes, it has weak profiles accuracy. - The service is not available in case of updating behavior profiles. - Not able to issue alerts in the right time.
Stateful protocol analysis (SPA)	- Knows and analyzes the protocol states. - Able to detect unexpected commands.	- Consumes more resources to trace protocol state. - Unable to detect attacks that resemble protocol behavior. - Compatibility problem with some operating systems and applications.

2.3.3.3 Stateful Protocol Analysis (SPA)

In these methods, the stateful protocol analysis tracing the profiles of expected protocol activity in each state and compares it with the vendor's original protocol. If the result does not match, the SPA will raise an alert and report it to the network administrator.

Table 2.1 shows the advantages and disadvantages of the IDS methodology.

2.4 CLASSIFICATION OF COMPUTER AND NETWORK ATTACKS

Due to the increase of technology use in different business areas, especially in computer systems and network, the number of vulnerabilities and attacks increases. Therefore, it is very important to know which type of attacks you may face, to help you prepare solutions according to the expected attacks. The following is a list of attack types.

2.4.1 ATTACK TYPE

The most common attacks include:

2.4.1.1 DoS Attack

DoS stands for denial of service. This type of attack is achieved by making the organization's system or network services unavailable for the user, using several technical methods such as flooding the victim with unwanted traffic or sending wrong information that triggers a system crash. In both ways, the result is that the system or network service is not available for use. DoS attacks use two methods to attack victims: flooding services and crashing.

Vulnerability and Attack Detection Techniques

Flooding attacks work by compromising the system because too much traffic goes to the server's buffer, which results in the server's response slowing down, for example, an ICMP flood and SYN flood.

An advanced feature of the DoS attack is called Distributed Denial of Services (DDoS) attacks. DDoS attacks use more than one compromised system, called zombies, to synchronize DoS attacks simultaneously to a single victim. The main difference between Dos and DDos attacks is that DDoS attacks are performed from different locations.

2.4.1.2 Probing Attacks

This type of attack is implemented by collecting information about a victim by scanning the network to identify the victim's IP address, operating system, and other useful applications in the victim's system. Then, the hacker exploits the vulnerability to perform attacks. The most common tools using probing attacks are the IP swap, port scan, and map [2].

2.4.1.2.1 Compromise

Compromise attacks are performed by gaining access privileges to a host. The hacker gains access through a known vulnerability or weak security points in the system. This attack's most common types are called user to root (U2R) and remote to local (R2L)

2.4.1.2.2 Worms, Viruses, and Trojans

Worms, viruses, and Trojans are developed using a program that can duplicate itself in the hosted machine and be distributed through the network.

2.4.1.3 Number of Network Connections Required by the Attacks

Attacks can be categorized by the number of network connections required to achieve them.

The most common attacks are:

- Attacks that require multiple network connections. The most common examples are probing and DoS.
- Attacks that require a single network connection, for example, buffer overflow.

2.5 SOURCE OF THE ATTACK

Most hackers start to attack from one location. Still, many location sources are needed to launch this attack, in the case of advanced DDoS attacks or any other plan attacks. Usually, these attacks target single victims or multiple victims. To detect this type of attack requires in-depth analysis of network data from different locations [5].

2.5.1 ENVIRONMENT

Attacks based on environment can be classified as:

2.5.1.1 Intrusion on Local Host Machine

This type of intrusion achieves on a specific host not necessarily connected to the network. It is very easy to detect this type of intrusion, because the attacker is associated with the user name. Analyzing system information (analyze system logs and command) can discover it.

2.5.1.2 Network Intrusion

Typically, this is done through the computer network, and usually, this type of intrusion comes from outside the organization. This type of intrusion can be detected by analyzing network traffic.

2.5.1.3 Intrusion in Wireless Network

This type of intrusion occurs in a computer connected via a wireless network. To detect this type of intrusion, you need to analyze wireless traffic stored in the wireless access point.

2.6 AUTOMATED LEVEL

This type depends on the level of an attack's automation, and it can be categorized as the following:

2.6.1 AUTOMATION ATTACKS

This type of attack typically uses automated tools to scan the network and probing to collect the needed information quickly. By using this tool, the work of attackers becomes very easy and even inexpert. Attackers can perform very high-risk attacks by using this tool.

2.6.2 SEMIAUTOMATED

This type of attack typically can be achieved by creating an automated script to scan the network and compromise the machine by installing attack code.

2.6.3 MANUAL ATTACKS

This type of attack is carried out manually; therefore, it needs more work and wide knowledge. This type of attack is not common, but it is a hazardous attack used only by expert people or organizations.

Vulnerability and Attack Detection Techniques

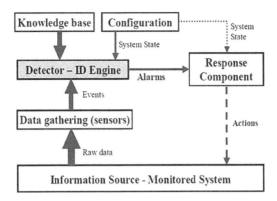

FIGURE 2.2 Components of intrusion detection system.

2.7 INTRUSION DETECTION SYSTEM COMPONENT

Many types of intrusion detection systems have been proposed by different researchers. They are technically different, but most intrusion detection systems consist of the following main components, as shown in Figure 2.2:

2.7.1 DATA GATHERING COMPONENT

This is a sensor device responsible for collecting data from the system that requires monitoring.

2.7.2 KNOWLEDGE-BASED COMPONENT

The database stores the information collected from the sensor in a preprocessed format, such as an attack's knowledge base, signature, and data filter.

2.7.3 CONFIGURATION COMPONENT

This is a device that provides information about the state of IDS.

2.7.4 RESPONSE COMPONENT

This is a device that is alerted when an intrusion is detected. This alert response can either be automated, or considered an active alert, or require human interaction, which is regarded as an inactive alert.

2.8 CONCLUSION

IDS technology is now considered one of the main components required to secure both small and large organizations. Therefore, IDS technology has improved dramatically to analyze real-time applications, sniff out malicious activity, and control high-speed networks with very complex traffic.

In this chapter, we provide an overview of the IDS and computer attacks. This overview is based on presenting the IDS methodology and approach, plus how the IDS interact to detect different types of computer and network attacks.

2.9 FUTURE WORK

IDS technology has a very promising future, and it is developing rapidly. More studies should be paid on the IDS issues in the virtual machine environment and secure internet cloud services in future work.

REFERENCES

[1] Kadam, P. U., & Deshmukh, M. (2014). Various approaches for intrusion detection system: An overview. *International Journal of Innovative Research in Computer and Communication Engineering*, 2(11), 6894–6902.

[2] Lazarevic, A., Kumar, V., & Srivastava, J. (2005). *Intrusion detection: A survey. In Managing Cyber Threats*. Springer, Boston, pp. 19–78.

[3] Depren, O., Topallar, M., Anarim, E., & Ciliz, M. K. (2005). An intelligent intrusion detection system (IDS) for anomaly and misuse detection in computer networks. *Expert Systems with Applications*, 29(4), 713–722.

[4] Jyothsna, V. V. R. P. V., Prasad, V. R., & Prasad, K. M. (2011). A review of anomaly based intrusion detection systems. *International Journal of Computer Applications*, 28(7), 26–35.

[5] Vinchurkar, D. P., & Reshamwala, A. (2012). A review of intrusion detection system using neural network and machine learning. *International Journal of Engineering Science and Innovative Technology*, 1(2), 54–63.

[6] Cintuglu, M. H., Mohammed, O. A., Akkaya, K., & Uluagac, A. S. (2016). A survey on smart grid cyber-physical system testbeds. *IEEE Communications Surveys & Tutorials*, 19(1), 446–464.

[7] Shone, N., Ngoc, T. N., Phai, V. D., & Shi, Q. (2018). A deep learning approach to network intrusion detection. *IEEE Transactions on Emerging Topics in Computational Intelligence*, 2(1), 41–50.

[8] Ferrag, M. A., Maglaras, L., Moschoyiannis, S., & Janicke, H. (2020). Deep learning for cybersecurity intrusion detection: Approaches, datasets, and comparative study. *Journal of Information Security and Applications* 50, 102419.

[9] Elhabashya, A. E., Wells, L. J., & Camelio, J. A. (2019), Cyber-physical security research efforts in manufacturing—A literature review, *Procedia Manufacturing*, 34, 921–931.

3 Digital Rights Management in a Computing Environment

Maram Bani Younes and Nameer N. El-Emam
Philadelphia University, Amman, Jordan

CONTENTS

3.1 Introduction to Ethics and Technoethics 27
    3.1.1 Definition of Ethics 27
    3.1.2 Definition of Technoethics 28
    3.1.3 Ethical Challenges in Technology 28
    3.1.4 Current Technoethics Issues 30
3.2 Cybersecurity and Its Applications 31
    3.2.1 Concepts of Cybersecurity 31
    3.2.2 Threats and Challenges of Cybersecurity 32
    3.2.3 Elements of Cybersecurity 33
    3.2.4 Cybersecurity Applications 34
3.3 Ethics for Cybersecurity Applications 36
    3.3.1 Privacy 36
    3.3.2 Freedom of Speech 37
    3.3.3 Intellectual Property Rights 38
    3.3.4 Legal Protections and Responsibility for Crimes 39
3.4 The Ethical Use of Machine Learning in Cybersecurity 39
3.5 Summary 40
References 40

3.1 INTRODUCTION TO ETHICS AND TECHNOETHICS

3.1.1 DEFINITION OF ETHICS

Human beings check the rules and regulations of any organization or community they join, to understand and follow them. This defines their rights and responsible tasks. Morals and ethics are internal principles that control the behavior of people in the absence of defined rules or surveillance situations [1]. Ethics constructs a system of moral rules that people are raised with, accordingly. It is usually inherited from religion, culture, or the surrounding nature. It affects the way people live and the decisions they make. People can distinguish what is right from what is wrong,

DOI: 10.1201/9781003145042-3

according to their ethics. The rights and responsibilities that control the relationships between people are all obtained from their ethics. The ways of obtaining rights and handling responsibilities reflect ethical principles, as well. There is no certain ethical answer to any situation or question. There is always a set of principles that could lead to several ethical answers regarding any question [2]. For example, one of the main issues regarding an ethical system is how people treat or think of each other. Each work environment defines a set of ethical rules that participants are required to follow. For example, there are ethical systems for the medical branch, where the privacy of patients should be completely protected. Moreover, there are ethical systems for courts and lawyers, where trust and secrecy should be guaranteed. In addition, there are ethical regulations for courts and lawyers, where trust and confidentiality must be guaranteed, while ensuring the existence of surveillance to detect unethical cases and report them to the official authorities.

3.1.2 Definition of Technoethics

One of the main fields that early on considered defining a specialized ethical system is computerized technology. This includes automated systems, communications, computerized systems, and the Internet. Technoethics are well known as the ethics of technology. However, technology is a tool [3], which clearly cannot possess morals or ethics. Technology is vital in our lives. Most of the difficult works that required a long processing time for humans in the past are finished automatically these day, quickly and without human intervention, using machines and technology. However, no emotions, feelings, or passion, either among the team workers or with the produced job, are expected. However, humans who are dealing with machines all the time and being served by tools and technology should not be converted into robots. Technology exists to facilitate the life of people and to help them, but not to kill their humanity. "The more we want to learn about technology, the more we need to understand about being human," as Stephen Petrina put it [4]. Technoethics could be simply defined as the ethics and principles that human possess when using or producing technology. However, several domains of knowledge have been considered in the field of technoethics research area, including social science and technology features. The detailed nature of the investigated technology usually affects its required principle/principles, to make sure that the developed technology appropriately enters our lives. Access rights, existential risk, health, over-automation, privacy, security, and terms of service are the main areas of technoethics [5]. The challenges and current issues of technoethics are investigated in the following two sections.

3.1.3 Ethical Challenges in Technology

Here we investigate major challenges for technoethics to address:

- **Diversity**

  The Internet and communication systems have connected the entire world together. Some companies and organizations are virtually created, over multiple countries and continents [6]. People in different countries possess different beliefs and different moral systems, which may conflict in some scenarios. Setting a neutral global ethical system that controls the behavior of employees and participants all over the world is a challenging mission.

- **Dependability**

Setting an ethical system that controls the usability of all technology types is another challenging mission. This is due to the fact that each kind of technology has its own requirements and challenges [7]. For example, the ethical rule that could be set for using autonomous vehicles might not be applicable for using smart-phones or using social media. Each branch of technology requires an ethical system that handles its specialized details.

- **Anonymous Users**

The ability to use the Internet anonymously introduces the challenge that people can deny actions over the network. This damages the entire surveillance system; some users of a technology will not hesitate to carry out unethical actions, if they know there are no responsibilities that will result. Moreover, some users can set the parameters over a connecting network to involve other innocent users in an unethical action and getting the associated penalty and embarrassment. Proving the identity of a real user of the Internet is a challenging process [8]. This challenge can be catastrophic for Internet of Things applications, where victims can be set up for a serious crime or when attackers could reach the setting of air conditioners in our homes.

- **Trust Management**

This represents social trust, which is usually used to help in the decision-making process [9]. It is a real challenge to achieve trust and let objects alone make decisions that affect major aspects of our lives. For instance, the machines that decide and inject doses of medicine for patients, based on regular blind tests without any doctor interference, need a long time to gain the trust of a society. Reputation and time are serious challenges, since developments in the field of technology change very quickly.

- **Privacy**

Various devices and equipment are installed all over private homes daily. These include camcorders, global positioning systems (GPS), and smart devices. Moreover, we carry smartphones and smart watches almost everywhere, and various applications are installed on them. The applications installed ask many private questions claiming to identify the user [10]. However, answers to these questions are shared with unwanted parties, in some cases. Besides, installation agreements contain points that allow remote access to the camera, photo gallery, or microphone, attacking the privacy of the user, in many scenarios.

3.1.4 Current Technoethics Issues

Here we select some issues in technoethics related to information technology systems. We investigate the controversial ethical visions regarding their services that create issues for technoethics.

- **Copyright**

  Digital copyright is becoming more difficult to protect over the interconnected global world. This is because people can obtain any software, database, or service easily over the Internet. Hackers dedicate their time to crack the security requirements of the produced software and release it to users, either to try it or use it in a restricted fashion. These versions are illegal, and the users could be threatened with judicial accountability. Copyright is a debated topic, since it depends on the nature of the product, the producer, and the country's rules regarding copyrights [11]. Moreover, even legal users who follow copyright procedures could develop a new service or application that uses an existing one. The moral conflict in this scenario is among people who see that copyrights protect the content of the released product and other people who accept mashed-up products.

- **Cybercriminality**

  Cybercrime, also known as computer-oriented crime, is defined as any crime that involves computers and networks. Indeed, hackers and criminals have exploited vulnerabilities in modern technologies to commit criminal and prohibited activities [12]. The anonymity and facilities over computer networks and the Internet have contributed to the rapid growth of cybercrime activities. However, laws in several countries have not considered punishing or stopping those criminals.

- **Global Positioning System (GPS) and Drones Applications**

  Global positioning systems have been installed in various devices and equipment recently. The main function of this technology is to track objects or geolocate targeted areas. Companies and businesses have installed this technology on their vehicles, cell phones, or valuable devices. This helps to ensure accurate and remote surveillance and track employees and property. People driving and moving on a daily basis can easily find a targeted destination using this technology. Moreover, drones or unmanned aerial vehicles (UAVs) fly over cities. Military and government organizations sometimes use them to guarantee the safety of their citizens. Photographers and game lovers have enjoyed their time using drones to explore areas that are hard to see [13]. These technologies can be helpful and enable the government to provide real-time help in emergencies. However, misusing them opens the door for very serious threats that invade people's privacy, even inside their homes. Drones can be easily used to spy on anyone, and with GPS tracking systems, people cannot hide anymore.

Digital Rights Management

3.2 CYBERSECURITY AND ITS APPLICATIONS

3.2.1 Concepts of Cybersecurity

The concept of cybersecurity is connected to the CIA security model [14]. This is a triangular model with three components: confidentiality, integrity, and availability (CIA). The confidentiality element aims to prevent unauthorized disclosure of data, which only authorized people can read and explore. Integrity prevents unauthorized modification of data, where everybody can read and see the data, but only authorized people can write and modify its content. Availability prevents unauthorized withholding of data, so no one can remove it and unauthorized people are prevented from exploring it. Authorized people are designated by the system's access control policy. Figure 3.1 graphically illustrates the CIA security model.

People use authentication to prove they are authorized to access the requested data. Authentication is the process of proving the identity of a computer user; it is achieved through a certain factor that the user knows (e.g., a password), owns (e.g., an ID card), or does (i.e., a signature) [15].

Cybersecurity is a subset of information security that focuses on protecting the confidentiality, integrity, and availability of digital data. However, information security protects both physical and digital data. Digital data is founded in one of three ways: stored on memory and servers, processed in CPUs, or transmitted over the networks. Data transmitted over the network is the most threatened type, because it can be more easily reached, processed, or held. Network security is developed to guarantee the security of all data sent or received by any device over a computer network and is considered a subset of cybersecurity [16]. Figure 3.2 graphically shows the hierarchical relationships between information security, cybersecurity, and network security.

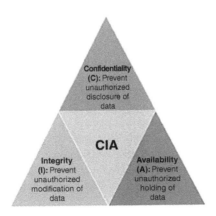

FIGURE 3.1 CIA triangle security model.

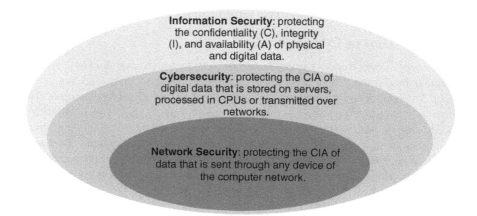

FIGURE 3.2 Hierarchical relation of information security, cybersecurity and network security.

3.2.2 THREATS AND CHALLENGES OF CYBERSECURITY

Here we discuss the main threats to cybersecurity, investigating the most common attacking techniques that target secure systems. Then, we illustrate recent and sophisticated challenges that face cybersecurity.

Threats to Cybersecurity

- Malware: A kind of malicious software that is represented by programs that aim to damage or gain unauthorized access to a system [17]. The most common and recent types of malware are:
    1. Viruses: replicate themselves by infecting other files on the same computer. Infected files involve the code of the malicious virus.
    2. Worms: aim to spread over the connecting networks. They replicate themselves and infect programs on other computers.
    3. Spyware: gathers information regarding the user of the computer, without their knowledge or acceptance.
    4. Adware: repeatedly displays unwanted and unexpected advertisements and interrupts the user.
    5. Trojans: hides its content and intent under an appealing fake application.
    6. Ransomware: encrypts important files or changes the passwords of legal users and blackmails them into paying ransom.
- Social engineering: A set of malicious activities that require human interaction to succeed. Attackers contact targeted victims, to convince them to provide information or set vulnerabilities in their secure system [18]. Malware is usually spread through these activities. The most common forms of social engineering attack techniques are:
    1. Baiting: false promises are used to lead victims into a trap that records their personal information or injects malware into their computer.
    2. Scareware: false alarms and threats push victims to take actions that allow their system to be infected with malware.

Digital Rights Management

3. Pretexting: a series of lies is carefully constructed to gain the trust of the victim and then obtain critical targeted information.
4. Phishing: masquerades as an authoritative person in an electronic communication, to gain sensitive or valuable information.

Challenges of Cybersecurity

- Anonymity: most actions over computer networks are taken by unknown people. The identity of users is not requested for most actions. Impersonation is also easier to achieve over a computer network because of the authentication factor, which can be fraudulent. Being anonymous and not responsible for the actions taken encourages attackers to damage systems and steal information [19]. Securing authentication factors and convincing anonymous people to behave over a network is a challenging process.
- Involving people: social engineering attacks rely on the reaction of authorized people in a targeted system [20]. The people are convinced differently, and they learn differently. Educating authorized people to protect the security of their system is a challenging process, especially since attackers keep changing their techniques.
- Automation and AI expansion: machine learning and automation have shown a promising future that serves humans with a smart living environment [21]. However, the database gathered is vital to these applications. Any manipulation of the database could cause catastrophic effects on the intelligent application.
- IoT Threats: people are not interacting in the environment of the Internet of things [22]. While this reduces the percentage of human error, it also introduces a serious challenge, in that data controls the life of humans. All the information in the system could be comprised, and attackers could prevent people from using their intelligent tools properly.

3.2.3 ELEMENTS OF CYBERSECURITY

Cybersecurity consists of six key elements that protect digital data [23], which exists in three main manners: stored, processed, or transmitted. So, a separate element handles each manner. Stored data is tackled by information security elements. Processed data is tackled by application security and operational security elements. Transmitted data is processed by network security. Moreover, disaster recovery and end-user education elements are used to retrieve or protect on-demand data. Figure 3.3 illustrates the main elements of cybersecurity. A brief discussion regarding each element is introduced in the rest of this section; meanwhile, the application security element is discussed in more detail in Section 3.2.4.

- **Network security**
  The procedure and policies to prevent intruders over computer networks. This includes checking the authentication of each user [24]. Then, firewalls check the policies and allowed services for that user [25]. Finally, antiviruses check the content of any installed program, to detect malicious

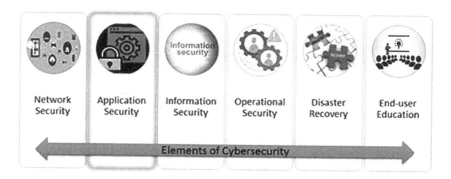

FIGURE 3.3 Elements of cybersecurity.

software [26]. Network security aims to protect the data transmitted among all devices on a computer network.

- **Application security**

  Aims to keep software and developed applications free of threats. It works at each phase of the software development life cycle. The compromised applications that attack data it claims to protect are targeted by this element to be detected and reported.

- **Information security**

  Guards sensitive information and protect its confidentiality, integrity, and availability. Protecting business records, employee data, and customer data are essential to any organization.

- **Operational security**

  Includes processes and decisions for handling and protecting data. It aims to protect the functionality of the organizations and eliminate all vulnerabilities to access sensitive data.

- **Disaster recovery**

  Plans that organizations set up to respond to any incident or data loss due to a security breach. Disaster recovery policies determine how can the organization affected can restore its operations and information quickly.

- **End-User Education**

Focuses on teaching legal users of the system to deal with suspicious scenarios. It aims to mitigate the effects of social engineering attacks.

3.2.4 CYBERSECURITY APPLICATIONS

A computer application is a software program that runs on a computer to provide a certain service to an end-user [28]. Email, web browsers, word processors, and

Digital Rights Management

35

calculators are all examples of installed applications that users need daily. In each field of study or work, there are several dedicated applications used to facilitate the mission of students or employees. For example, computer-based education (CBE) applications help students to follow their courses and emphasize their own studies. Meanwhile, applications have been developed to assist people in the health care field, such as a lab-diagnostic application, which is used to document all tests and reports done in a laboratory.

Moreover, regardless of the field of work, each organization introduces specialized applications to serve its employees and customers. Due to the Internet and communication evolutions, web and mobile applications are becoming widely used and developed. In these applications, the same functionalities are provided to the user by running the program on remote web servers or intelligent mobile devices.

These applications could be targeted by attackers and intruders in various ways, with various objectives. Applications could be developed and advertised to each system aiming to steal, hijack, or manipulate data and information that it proposes to process or protect. Hackers also block legal applications and prevent users from executing or reaching them. Each application has different vulnerabilities that could be exploited, in order to attack the system. Developing a secure application requires including a security consultant, who assets and mitigates vulnerabilities at each phase of the software development life cycle. Moreover, updated algorithms and security approaches are required, to keep protecting applications after they are released. The security of web and mobile applications suffers from multiple challenges and requires special settings, mainly when transferring critical data over the Internet. A brief discussion regarding the security of these types of applications follows.

- **Secure mobile applications**

    Running applications over mobile devices introduces several vulnerabilities, such as reverse engineering, client-side injection, insecure data storage, and insufficient encryption. The recent expansion in running applications over mobile devices has been strongly supported by developing secure protocols and mechanisms that oppose these vulnerabilities. Using the obfuscation code technique makes the code unreadable and then prevents a reverse-engineering attack [29]. In the latter attack, hackers create clones to the targeted application that guide users to open the way for a malicious attack. Validating input data also prevents injection attempts on the server. Moreover, encrypting all stored data and using multi-factors for authentication enhances the security setting and encourages users to install and run mobile applications.

- **Secure web applications**

Running applications remotely on a server introduces vulnerabilities that attackers could exploit. The cross-site script, SQL injection, denial of service, buffer over flow, and memory corruption are attacks target a web-hosting server. These techniques aim to manipulate a proposed application or prevent it [30]. More sophisticated attacks can utilize the botnet technique to achieve a large-scale network disruption, the distributed denial of services (DDOS). Web application firewalls (WAF) are used as barriers to stop attackers and prevent malicious access to the targeted servers.

3.3 ETHICS FOR CYBERSECURITY APPLICATIONS

The ethical principles and rules that control the behavior of individuals in any society enhance the functionality and profits of an adapted business and protect human and employee rights. The reputation of any business or organization can help to gain the respect and trust of its customers and other competing brands. Highly advanced technology and communications have changed the shape of organizations and working environments to cover continents. Services and functionalities are provided through mobile and web applications to end users. Ethical systems that can face serious challenges and issues are required, to keep the reputation of these applications and gain the trust of customers.

Surveillance and security control are limited over these applications. However, ethical requirements for cybersecurity applications are essential, to maintain the growth and continuity of these applications. In this section, we discuss four main ethical requirements of cybersecurity applications.

3.3.1 PRIVACY

Over cyberspace, a virtual world has been completely built that cannot be separated from our daily real world. However, protecting the privacy of users and the data collected over websites is a difficult mission. People need to create a user record providing information before being able to access any website service. Providing information on each website makes the stalking process easy for targeted individuals. This scares users and prevents the high usability of web-based applications.

Websites are responsible for protecting the privacy of their users, to gain their trust and continuous operations. First, sharing the data collected with any other companies or sources is ethically prohibited, even when this data is general or public. Second, the website should protect their servers and eliminate vulnerabilities that allow intruders to reach or manipulate the collected data. Cross-site scripting, SQL-injection, and XML-injection are the most popular client-side attacks that aim to reach databases of collected data. Third, critical data provided, such as bank accounts or medical specifications, require extra protection for user privacy and to avoid legal penalties. Encrypting this collected data provides a partial solution, while requiring more processing time and extra memory [31].

On the other hand, mobile applications require installing the software of a targeted application on the mobile device and running it. Some mobile applications can access the camera, microphone, GPS location, or photo gallery of a smartphone. This problem is more severe, since mobile applications can collect all activities, words, behaviors, and even feelings of the people that use the phone and install these applications. The aim is to provide ethical protection for users that enable them to install and use these applications without threatening their privacy. First, each of these applications should notify or request the ability to use the data or other private applications of the mobile phone. Second, the application cannot share any of the user data with other sources without the approval of users. Figure 3.4 graphically illustrates the main attacks on the privacy of users using the web and mobile applications.

Digital Rights Management

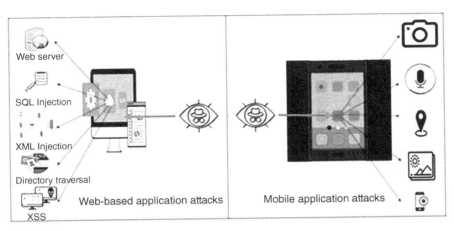

FIGURE 3.4 Privacy attacks using web and mobile applications.

3.3.2 FREEDOM OF SPEECH

Social media applications such as Twitter, Facebook, and Instagram connect people over the entire world. News, advertisements, and rumors spread over the social network faster than expected. Friendships and relationship following are created or discarded over these applications, based on the similarities and conflict of interest or opinions among the people involved. Daily sharing of posts regarding certain issues affect the taste and opinions of followers. Most users create accounts that can lead to their real identities, sharing their real names, addresses, jobs, or ages. However, some users provide misleading names and details, to use social media without revealing their true identity. Some users provide their real identity without giving too many details.

Social media were initially introduced to enhance human socialization over virtual cyberspace and keep people connected, regardless of distance [32]. However, they are now widely used for advertising and political orientation. Most recent businesses advertise and market their products and services over social media. Even customers find it more convenient to shop online, while exploring their favorite social page. Politicians and celebrities have found social media an easy way to reach more people and gain more attention or fans, while posting their achievements, criticisms, or supporting directions. Several people have stopped surfing journalism or news pages and now feel satisfied with the information on social pages.

Social media have given the power of broadcasting to individuals, which was previously limited to newspapers and television. Users of the media are free to choose their friends and post their opinions, feelings, adventures, and experiences on their pages. However, ethics should control the contents of the posts shared on social media. They should be free from offensive content; users are not free to insult or attack other people or their beliefs, and belonging publicly. Political and religious opinions should be posted with caution, especially by popular and effective users. Most important, all posts should avoid obscenity, racism, or criminal prompting.

Global censorship is required over cyberspace, and especially over the material posted on social media. Anonymity protects users and makes it hard to reach, question, or punish them. Reporting unethical account usually leads to blocking and closure of an account. Censorship is a mission for the entire community and users over cyberspace, because the huge domain of users and daily posts cannot be controlled completely by any center.

3.3.3 Intellectual Property Rights

Intellectual property rights include copyright, neighboring rights, patent, design, trademark and trade secrets, among many others [33], as shown in Figure 3.5. Art, poetry, and individual creative works such as songs, movies, or developed software need to be protected from illegal copying or use. Only copyright holders should use these products and develop them. Patents occur in the industrial branch, forcing companies and factories to buy a idea they aim to manufacture and use. Moreover, the business is protected by trademarks and trade secrets.

In cyberspace and over Internet applications, copyrights are the most important applicable rights to protect. This is due to the simple producer of recording, copying, and transferring small and large electronic products over these advanced technologies. Copyright and similar rights exist to encourage people and enhance the rewards they obtain from the intangible creations of their intellect.

Sharing a video, article, poem, or song on the Internet is a daily practice that users follow. The first moral aspect that users should check before sharing these kinds of content is to include the source and give credit to the creator. The second important aspect that users need to check before sharing content is the details of copyrights,

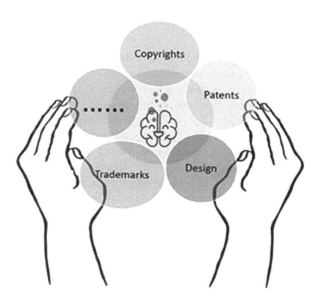

FIGURE 3.5 Protecting intellectual property rights.

Digital Rights Management   **39**

because owners may have concerns regarding distributing or displaying their products.

Software and developed applications are categorized as art and protected by copyright [1]. Some software is developed as free or open-source, meaning that it can be used and developed legally by any person, without asking for permission. However, other software is only available to copyright holders who buy the license to use it. Crackers and pirates have targeted the latter type of software, aiming to obtain illegal versions without paying the owners. Regulations have been set to determine the penalties and sanctions for these attackers.

3.3.4   LEGAL PROTECTIONS AND RESPONSIBILITY FOR CRIMES

Cybercrime and hacking activities over the Internet have become dangerous issues that threaten human societies. These crimes are intentionally committed by using computer networks and devices. Identity theft, phishing, cyber terrorism, child pornography trafficking, and other cybercrimes are detected daily and reported over the Internet. On the other hand, people, organizations, and governments regularly use the Internet for their work and activities. This requires a level of protection that assures users can proceed on the Internet without becoming victims of such attacks.

Regulations and legal penalties have been set to punish and deter cybercriminals [34]. These penalties are assigned according to the amount of damage or importance of the stolen information. This is usually affected by the rules followed in a country or among its judges. Digital actions usually cross borders, while the law is different in each country. Thus, an action that is considered legal in the home country might be illegal in another country. Expanding a certain business over multiple countries should adopt and comply with the different laws of the countries involved.

The huge usage and anonymity of the Internet make it difficult to trace the source of criminal activities. Moreover, identity theft and impersonation could lead to innocent users, causing them to be accused of the crime. Users need to prove their identities to gain trust and acceptance; anonymous sources are untrusted sources. Victims need evidence and proof to be able to sue the hackers responsible for an attack.

3.4   THE ETHICAL USE OF MACHINE LEARNING IN CYBERSECURITY

Machine-learning-based algorithms and techniques have been developed to intelligently solve computer problems. Machine-learning-based algorithms have a proven evolution in the field of decision-making, where less human effort and involvement are required. Moreover, extra sophisticated, powerful, and quick solutions are produced. Most recent computerized applications are developed according to one or more machine-learning techniques. The future of robotic and autonomous tools that serve humanity all follow the essential principles of machine learning.

Hackers, intruders, and attackers have utilized the power of machine learning algorithms to make their activities their activities and reach the targeted information, system, or network [35,36]. Detecting these malicious activities and defending against them are more complicated processes, compared to traditional attacking techniques [37]. The fact that details regarding the implementation of the machine learning-based techniques are hidden and unknown makes any attack activity strong and hard to be defeat completely.

On the other hand, security developers have proposed several defending algorithms and secure protocols using machine-learning-based techniques and principles [38]. These techniques work well, in terms of protecting the targeted system. However, the developers cannot guarantee the behavior of these techniques in all of the challenging scenarios involving ethical risk scenarios, due to a lack of details regarding how the mechanism has reached the concluding results [39]. Cybersecurity applications should consider the ethical risks involved in each scenario and the details of the machine-learning technique used, to make it more understandable and transparent [40].

3.5 SUMMARY

This chapter investigates the ethical issues for cybersecurity applications. First, we define the related concepts of ethics and how it appears in the field of technology. Then, cybersecurity is discussed, illustrating its threats, challenges, and main elements. Cybersecurity applications are investigated in more detail, where web-based and mobile-based applications are the main applications studied. The ethics of these applications are reviewed, in light of the privacy of users, freedom of speech, intellectual property, and responsibility for crimes. Finally, the ethical use of machine learning in cybersecurity is briefly investigated.

REFERENCES

1. Myers, Glenford J., Tom Badgett, Todd M. Thomas, and Corey Sandler. *The art of software testing*. Vol. 2. Chichester: John Wiley and Sons, 2004.
2. Januszewski, Al, and Michael Molenda, eds. *Educational technology: A definition with commentary*. Routledge, 2013.
3. Boudreau, Marie-Claude, and Daniel Robey. "Enacting integrated information technology: A human agency perspective." *Organization Science* 16, no. 1 (2005): 3–18.
4. Luppicini, Rocci, and Arthur So. "A technoethical review of commercial drone use in the context of governance, ethics, and privacy." *Technology in Society* 46 (2016): 109–119.
5. Klapper, Leora, Raphael Amit, Mauro F. GuillÃl'n, and Juan Manuel Quesada. *Entrepreneurship and firm formation across countries*. Washington, DC: The World Bank, 2007.
6. Goleva, Rossitza, Dimitar Atamian, Seferin Mirtchev, Desislava Dimitrova, Lubina Grigorova, Rosen Rangelov, and Aneliya Ivanova. "Traffic analyses and measurements: Technological dependability." In George Mastorakis and Constandinos X. Mavromoustakis (Eds.), *Resource Management of Mobile Cloud Computing Networks and Environments*, pp. 122–173. IGI Global, 2015.
7. Palme, Jacob, and Mikael Berglund. "Anonymity on the Internet." (2002). Retrieved August 15, 2009, from http://people.dsv.su.se/~jpalme/society/anonymity.html
8. Blaze, Matt, Joan Feigenbaum, and Jack Lacy. "*Decentralized trust management*." In *Proceedings 1996 IEEE Symposium on Security and Privacy*, pp. 164–173. Oakland, CA, USA: IEEE, 1996.

9. Joinson, Adam N., and Carina B. Paine. "Self-disclosure, privacy and the Internet." *The Oxford handbook of Internet Psychology* 2374252 (2007).
10. Stokes, Simon. *Digital copyright: Law and practice.* Oxford: Bloomsbury Publishing, 2019.
11. Corbet, Shaen, Douglas J. Cumming, Brian M. Lucey, Maurice Peat, and Samuel A. Vigne. "The destabilising effects of cryptocurrency cybercriminality." *Economics Letters* 191 (2020): 108741.
12. Solodov, Alexander, Adam Williams, Sara Al Hanaei, and Braden Goddard. "Analyzing the threat of unmanned aerial vehicles (UAV) to nuclear facilities." *Security Journal* 31, no. 1 (2018): 305–324.
13. Samonas, Spyridon, and David Coss. "The CIA strikes back: redefining confidentiality, integrity and availability in security." *Journal of Information System Security* 10, no. 3 (2014): 21–45.
14. Raj, Thanigaivel Ashwin. "Authentication process for value transfer machine." U.S. Patent Application 13/593,245, filed August 29, 2013.
15. Von Solms, Rossouw, and Johan Van Niekerk. "From information security to cyber security." *Computers & Security* 38 (2013): 97–102.
16. Aycock, John. *Computer viruses and malware.* Vol. 22. Springer Science and Business Media, 2006.
17. Salahdine, Fatima, and Naima Kaabouch. "Social engineering attacks: A survey." *Future Internet* 11, no. 4 (2019): 89.
18. Berthold, Oliver, Hannes Federrath, and Marit Köhntopp. *"Project anonymity and unobservability in the internet."* In *Proceedings of the Tenth Conference on Computers, Freedom and Privacy: Challenging the Assumptions,* pp. 57–65. Toronto, Canada: ACM, 2000.
19. Feng, Shuo, Peyman Setoodeh, and Simon Haykin. "Smart home: Cognitive interactive people-centric Internet of Things." *IEEE Communications Magazine* 55, no. 2 (2017): 3439.
20. Zekri, Marwane, Said El Kafhali, Noureddine Aboutabit, and Youssef Saadi. *"DDoS attack detection using machine learning techniques in cloud computing environments."* In *2017 3rd International Conference of Cloud Computing Technologies and Applications (CloudTech),* pp. 1–7. Rabat, Morocco: IEEE, 2017.
21. Abomhara, Mohamed. "Cyber security and the Internet of Things: Vulnerabilities, threats, intruders and attacks." *Journal of Cyber Security and Mobility* 4, no. 1 (2015): 65–88.
22. Newmeyer, Kevin P. "Elements of national cybersecurity strategy for developing nations." *National Cybersecurity Institute Journal* 1, no. 3 (2015): 9–19.
23. Brickell, Ernie, and Wesley Deklotz. "Identity authentication portfolio system." U.S. Patent Application 10/017,835, filed June 19, 2003.
24. Morgan, Dennis, Alexandru Gavrilescu, Jonathan L. Burstein, Art Shelest, and David LeBlanc. "Method of assisting an application to traverse a firewall." U.S. Patent 7,559,082, issued July 7, 2009.
25. Dyer, Allen R. *Ethics and psychiatry: Toward professional definition.* Washington, DC: American Psychiatric Association, 1988.
26. Sparks, John R., and Yue Pan. "Ethical judgments in business ethics research: Definition, and research agenda." *Journal of Business Ethics* 91, no. 3 (2010): 405–418.
27. Lai, Everett, Patrick Gardner, and John Meade. "Method for mitigating false positive generation in antivirus software." U.S. Patent 8,087,086, issued December 27, 2011.
28. Kozma, Robert, Joao Luis G. Rosa, and Denis R. M. Piazentin. *"Cognitive clustering algorithm for efficient cybersecurity applications."* In *The 2013 International Joint Conference on Neural Networks (IJCNN),* pp. 1–8. IEEE, 2013.

29. Graa, Mariem. "Hybrid code analysis to detect confidentiality violations in android system." PhD Dissertation, Université de Rennes, 2014.
30. Chong, Stephen, Jed Liu, Andrew C. Myers, Xin Qi, Krishnaprasad Vikram, Lantian Zheng, and Xin Zheng. "Secure web applications via automatic partitioning." *ACM SIGOPS Operating Systems Review* 41, no. 6 (2007): 31–44.
31. Li, Yibin, Keke Gai, Longfei Qiu, Meikang Qiu, and Hui Zhao. "Intelligent cryptography approach for secure distributed big data storage in cloud computing." *Information Sciences* 387 (2017): 103–115.
32. Correa, Teresa, Amber Willard Hinsley, and Homero Gil De Zuniga. "Who interacts on the Web?: The intersection of users personality and social media use." *Computers in Human Behavior* 26, no. 2 (2010): 247–253.
33. Yu, Peter K. "Intellectual property and the information ecosystem." *Michigan State Law Review* 2005, no. 1 (2005): 1–20.
34. Hui, Kai-Lung, Seung Hyun Kim, and Qiu-Hong Wang. "Cybercrime deterrence and international legislation: Evidence from distributed denial of service attacks." *MIS Quarterly* 41, no. 2 (2017): 497.
35. Tuan, Nguyen Ngoc, Pham Huy Hung, Nguyen Danh Nghia, Nguyen Van Tho, Trung Van Phan, and Nguyen Huu Thanh. "A DDoS Attack Mitigation Scheme in ISP Networks Using Machine Learning Based on SDN." *Electronics* 9, no. 3 (2020): 413.
36. Alqahtani, Hamed, Iqbal H. Sarker, Asra Kalim, Syed Md Minhaz Hossain, Sheikh Ikhlaq, and Sohrab Hossain. "*Cyber Intrusion Detection Using Machine Learning Classification Techniques*." In *International Conference on Computing Science, Communication and Security*, pp. 121–131. Gujarat, India: Springer, Singapore, 2020.
37. Verma, Abhishek, and Virender Ranga. "Machine learning based intrusion detection systems for IoT applications." *Wireless Personal Communications* 111, no. 4 (2020): 2287–2310.
38. Sarker, Iqbal H., Yoosef B. Abushark, Fawaz Alsolami, and Asif Irshad Khan. "IntruDTree: A Machine Learning Based Cyber Security Intrusion Detection Model." *Symmetry* 12, no. 5 (2020): 754.
39. Sarker, Iqbal H., A. S. M. Kayes, Shahriar Badsha, Hamed Alqahtani, Paul Watters, and Alex Ng. "Cybersecurity data science: an overview from machine learning perspective." *Journal of Big Data* 7, no. 1 (2020): 1–29.
40. Einstein, A., B. Podolsky, and N. Rosen, "An quantum-mechanical description of physical reality be considered complete?" *Physics Review* 47 (2020; 1935): 777–780.

4 Trade-Offs and Vulnerabilities in IoT and Secure Cloud Computing

Suman De
SAP Labs India Pvt. Ltd., India

CONTENTS

4.1	Introduction	44
4.2	History of IoT and Cloud Vulnerabilities	45
4.3	Literature Survey	46
	4.3.1 Recent Works in Cloud Computing	46
	4.3.2 Recent Works in Internet-of-Things	48
4.4	IoT and Similar Advancements	49
4.5	Risks and Breaches of IoT devices	51
	4.5.1 Verification	52
	4.5.2 Cryptography	52
	4.5.3 Modifying Techniques	52
	4.5.4 Physical Permit	52
	4.5.5 System Control	53
4.6	Cloud Computing	53
4.7	Vulnerabilities with Cloud Offerings	55
	4.7.1 Lesser Clarity with Control	55
	4.7.2 On-Demand Self-Service	56
	4.7.3 Worldwide Controlling APIs	56
	4.7.4 Multi-Tenant Feature	56
	4.7.5 Information Removal	56
	4.7.6 Stolen User Details	57
	4.7.7 Supplier Commitment	57
	4.7.8 Higher Complexity	57
	4.7.9 Insider Abuse	58
	4.7.10 Lost Information	58
	4.7.11 Provider Supply Chain	58
	4.7.12 Inadequate Due Perseverance	59

DOI: 10.1201/9781003145042-4

4.8	Secure Cloud Computing Techniques		59
	4.8.1	Infrastructure Security	59
		4.8.1.1 Physical Security	59
		4.8.1.2 Network Security	59
	4.8.2	Remote Security	60
	4.8.3	Host Security	60
	4.8.4	Security for Middlewares	60
		4.8.4.1 Containers	60
		4.8.4.2 Application Programming Interfaces (APIs)	60
		4.8.4.3 Databases	61
		4.8.4.4 Resource Management Platform	61
	4.8.5	Application System Security	61
	4.8.6	Data Security	61
4.9	Summary		62
References			62

4.1 INTRODUCTION

Digitization is at its height in the current world, with a global pandemic only accelerating the impact of a heavier virtual existence for every business and social convention. From editing a document through an offering from Microsoft, to attending an academic conference virtually using Skype, to managing one's business processes using SAP solutions, the world is dependent on solutions that touch both the cloud and the Internet-of-Things (IoT) aspects of technological advancements. The scenarios are endless, and the efforts are going towards realizing more and more processes in a safe and secure virtual environment that gives an experience to the end-user similar to what would have been with a physical presence. The global COVID-19 pandemic affecting the possibilities of physically conducting and completing business processes has given a stronger push for technology to focus even on topics such as augmented and virtual reality. A lot of these applications have a strong dependency on how and where data is stored in the cloud, thus raising concerns relating to data theft, mismanagement, etc.

We are also looking at a world where not just data is vast, but also the network with which the data is scattered and interconnected through a stream of devices. This chapter looks at the impacts that this overgrowing dependency can have under the subtopics of cloud computing and the IoT (see Figure 4.1). We explore these subjects with a focus on research and encourage the reader to identify areas with open challenges and find solutions that can be published on or taken forward as new research ideas.

The chapter is divided into subsections that cover a history of breaches that have affected some of the biggest organizations. It is followed by the latest works of researchers for the specific worlds of the cloud and IoT and explores how research enthusiasts, scholars, and industry professionals can utilize them for future research work and also to gain a better understanding of the topics. The chapter then explores the basics of vulnerabilities associated with the applications and concludes with a

Trade-Offs and Vulnerabilities

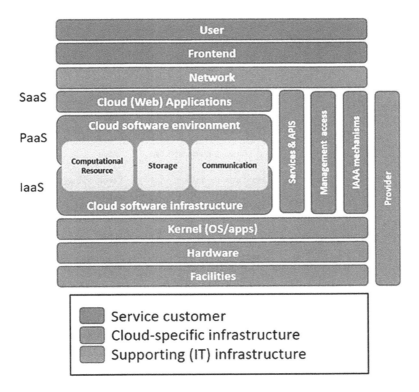

FIGURE 4.1 A typical cloud architecture, with various components that can be affected through associated vulnerabilities.

[Source: https://www.infoq.com/articles/iee-cloud-computing-vulnerabilities/].

section on techniques that involve secure cloud computing. Subsequent chapters cover IoT data-intensive secure systems and the cloud further, to provide an in-depth understanding of the subject matter.

4.2 HISTORY OF IOT AND CLOUD VULNERABILITIES

Although cloud computing and the IoT have seen a huge expansion in recent times, quite a few high-profile vulnerabilities have been seen in the last few years. In this section, we look at some famous incidents that had an adverse effect on the digital world. Companies realized that the cloud consists of couple of tips of choice with drawbacks taking everything into account. As indicated by an ongoing survey, security has been placed as the two primary features and best exam for distributed computing of Information Technology aces. The chapter concludes that attending to security issues is considered the best path for exploiting all the cloud has to bring to the table.

TABLE 4.1
High profile cloud/data breaches in history

Organization	Year	Issue	Remarks
Microsoft	2010	Breach in the Business Productivity Online Suite	Fixed with 2 hours and only a small number of users affected
Dropbox	2012	Hackers gained access to more than 65 million user accounts and almost 5 GB of data	Dropbox asked its user base to have a site-wide password reset
National Electoral Institute of Mexico	2016	More than 93 million voter database records were compromised	Poorly managed data over an insecure publicly available cloud server
LinkedIn	2012, 2016	User passwords are twice stolen, once for 6 million and again for 167 million users.	LinkedIn requested site-wide password change and introduced two-way authentication
Home Depot	2014	Attack on point-of-sale terminals affected 55 + million credit cards	Had to payout more than $100 million in settlements and consumer refunds
Apple iCloud	2014	iCloud Storage for celebrity profiles was compromised	Apple urged users to use strong passwords and brought in an alert system
Yahoo	2013	Accounts of more than 1 billion users were breached	History's largest data breach

4.3 LITERATURE SURVEY

Cloud computing and the IoT continue to be vital subjects for research in both the industrial and academic domains, including research completed concerning resource optimization, allocation, computation, etc., and, extensively, security. In this section, we look at the latest research done in the last couple of years for the topics under discussion and also form a base for research scholars and enthusiasts to take the sub-areas forward with further modifications and also address open problem statements that are still waiting to be solved. We start with a subsection on recent works focusing on cloud computing, before diving into similar research on the vulnerabilities seen in Internet-of-Things.

4.3.1 RECENT WORKS IN CLOUD COMPUTING

There have been a high number of papers relating to data maintenance, resource allocation, privacy, and similar topics that affect both the consumer and producers. We start by looking at the research work done by Narendra Mishra and R. K. Mishra, "Taxonomy & Analysis of Cloud Computing Vulnerabilities through Attack Vector, CVSS, and Complexity Parameter" (2019). The world is seeing an uncommon development in cloud-empowered administrations, which is further developing step-by-step, because of the progression and prerequisite of innovation. In any case, the distinguishing proof of weaknesses and their abuse in distributed computing will consistently be the significant test and worry for any distributed computing framework. To comprehend the difficulties and their results and further give relief methods for the weaknesses, the distinguishing proof of cloud-explicit weaknesses should be analyzed first; then, a point-by-point scientific classification must be situated. In this chapter, a few cloud-explicit recognized weaknesses are considered that can be recorded through ENISA CSA, NVD, and so forth. In like manner, a comprehensive scientific categorization for security weaknesses has been readied. The chapter discusses a thorough scientific categorization of cloud-explicit weaknesses, for boundaries such as attack vector, CVSS score, unpredictability, and so forth, and further looks at the contribution of specific investigations and moderation in cloud weaknesses. Conspiring in nomenclature of weaknesses for a successful path of heads, chiefs, purchasers, along with different partners in comprehending, recognizing, and tending to security chances.

In 2018, Rahul L. Pakirao and Varsha H. Patil researched "Security as a Service Model for Virtualization Vulnerabilities in Cloud Computing." The cloud is a cutting-edge worldview principally zeroing in on distant asset sharing for efficient registering. The cloud offers different support models and attempts to help the client, for a better asset for the executives. Because of distant asset sharing, not many security concerns emerge that need to be dealt with. In this chapter, we attempt to survey and investigate different security issues emerging because of the various parts of distributed computing. We second the contemplations put by different scientists and classify them extensively as issues distinguished by CSA, issues recognized because of the area of the information, issues acquired from systems administration, and all the more critically, issues that arise because of weaknesses in virtualization. Hypervisor weaknesses are a territory of worry, all things considered. Security as a help model is proposed for virtualization vulnerabilities.

Ching-Chiang Chen and Chia-Nan Wang (2019) present considerable research on preparing a secure model for cloud computing for vulnerability prevention. Cloud gives an incredible figuring stage is in the center, the security of the current loss of trust in distributed computing. This examination is dependent on the conventional ISO 27001 counterfeit review practice, as a waterproof divider in the cloud, with distributed computing security issues. In this chapter, this investigation depends on framework elements for a data security strategy to reenact the distributed computing security situation and delineate the distributed computing security anticipation system. Finally, the reenactment results show that the audit framework to anticipate the of server side's break farms in the system of distributed computing. The principal commitment has these three focuses: 1. The conventional online data security principles, through intelligent mechanized reviews, will bit-by-bit refresh the manual review mode. 2. Through the presentation of the WcsM waterproof divider idea in this examination, the conventional security insurance systems (Firewalls, IPSs, IDSs, and so on) will have more proactive measures to guarantee cloud data security. 3. The idea of decreasing the capital speculation by methods for mechanized reviews will additionally advance the significance of SMEs-based little and miniature endeavors.

Cloud computing research, as conducted by many authors, has found its way into both academics and industry and continues to be a driving factor for different aspects in the cloud are defined and optimized with the latest technological advancements.

4.3.2 Recent Works in Internet-of-Things

The connectivity of devices defines multiple aspects of digitization and is considered a critical area of research by researchers and industry, to making progress to better possibilities of virtual communication and the completion of business and consumer processes. In this section, we look at different research works related to vulnerabilities associated with the Internet-of-Things, which is also commonly addressed as Pervasive Computing.

First, we observe the work by Huan Wang, Zhanfang Chen, Jianping Zhao, Xiaoqiang Di, Dan Liu, "A Vulnerability Assessment Method in Industrial Internet of Things Based on Attack Graph and Maximum Flow" (2018). To comprehend the least attributed attack path assessment value and complicated pathfinding considering advanced Internet of Things scenarios, one shortcoming calculation methodology reliant concerning attack graphs along with the best flow gets presented. This strategy takes into account the factors affecting a specific attack, along with the association among network center points. The attack's danger is controlled by a common shortcoming scoring structure, which grows the attack way assessment degree. The best hardship stream depicts the attack way, evaluates the association's shortcoming by the most extraordinary adversity stream and mishap inundation, and addresses the shortcoming's significance. Avoiding a repetitive tally and obtaining the potential key shortcoming way rapidly, expanded road figuring is used to find the ideal attack path inside the overall way. The result shows for surveys are convincing, considering the shortcomings.

Next, we consider the paper titled, "A DDoS Attack Detection and Mitigation with Software-Defined Internet of Things Framework" (April 2018), by authors Da Yin, Lianming Zhang, and Kun Yang. Considering the reach involved with IoT use cases, security is turning out to be critical. An ongoing Distributed Denial-of-Service (DDoS) attack uncovered specific pervasiveness in weaknesses observed for Internet of Things scenarios This perspective gives an approach to securely oversee IoT devices. The article initially presents an overall structure in SD-IoT in view of the SDx worldview. The discussed system here comprises a regulator group consisting of SD-IoT regulators, SD-IoT controllers incorporated consisting of an IoT door, and IoT devices. The researchers proposed a calculation for identifying along with alleviating DDoS attacks utilizing the discussed SD-IoT system, along with the mentioned calculation, the cosine similitude through vectors for a specific bundle in message frequency that limit SD-IoT switching ports, which are utilized to decide if DDoS attacks happen in these scenarios. Finally, test results show that the proposed calculation has great execution, and the given structure is adjusted to reinforce specific security for an IoT scenario with diverse and weak devices.

We also look at an Intrusion Detection technique, through the work of Nour Moustafa, Benjamin Turnbull, and Kim-Kwang Raymond Choo, "An Ensemble Intrusion Detection Technique Based on Proposed Statistical Flow Features for Protecting Network Traffic of Internet of Things" (2019). For specific day-to-day activities, the IoT assumes, for certain situations, an undeniably critical function, linking objects around us into computerized networks. All in all, the IoT is the main

Trade-Offs and Vulnerabilities

thrust behind home mechanization, smart urban communities, present-day well-being frameworks, and progressed fabricating. This likewise improves the probability of digital dangers against IoT devices and administrations. Attackers may endeavor to abuse weaknesses in programs conventions, such as Message Queue Telemetry Transport (MQTT), Domain Name System (DNS), and Hyper Text Transfer Protocol (HTTP), which interface straightforwardly including the likes of background information base frameworks and customer worker applications to store the information of IoT administrations. The viable abuse of any one of these can achieve data spillage and security infiltration. In this chapter, a gathering obstruction recognizing confirmation methodology is introduced to decrease malevolent occasions, including botnet attacks opposing the DNS, HTTP, and MQTT types used for IoT associations. The latest quantifiable stream highlights are made using the seen dependents on an appraisal of their ordinary properties. By that point, one specific AdaBoost pack education philosophy has been made, utilizing 3 AI strategies, in a particular choice tree, Naive Bayes (NB), and phony neural association, to assess the impact of these highlights and separate risky occasions adequately. The UNSW-NB15 and NIMS botnet datasets with reenacted IoT sensors' information are utilized to eliminate the proposed joins and assess the outfit framework. The test results show that the proposed highlights have sensible qualities of the regular and malignant advancement utilizing specific core entropies and affiliation coefficient measures. In addition, a higher disclosure value at the organizational structure is discussed with a lower counterfeit positive rate and some of the top-level strategies.

This section focuses on the latest technical papers published in the fields of cloud computing and the IoT. The objective of this chapter is to create a clear path for readers to understand the nitty gritty of these topics, for usage and further research.

4.4 IOT AND SIMILAR ADVANCEMENTS

The IoT means that loads of genuine devices across this planet continue to be connected between the nodes of the internet, including every social activity, along with distributed data. In light of the presence of super-unassuming CPUs and the unavoidability of inaccessible affiliations, it's conceivable to connect anything, from something as small as a tablet to something as broad as a level, to the IoT. Interfacing all these various articles and adding sensors to them adds a degree of frontline comprehension that would be considered idiotic for devices, and empower them to offer predictable information with security concerns. The IoT scenario makes the exterior of our general ecological variables shrewder and more interactive, blending the mechanized and real universes.

The advantages for businesses using the IoT depend upon specific use, and in general provide adequate ability to examine with the highest degree of measures. The concern is which endeavors should move toward higher data about their things and their inward structures, and a more unmistakable ability to make changes in this way. Producers are adding sensors to portions of their things with the objective of conveying data to the source, based on the recorded performance. This can help companies, organizations, and service provides spot when a segment is apparently going to come

TABLE 4.2

IoT devices studied in groups, 2018 to 2020, global (installed base and billions in units)

Group	2018	2019	2020
Tools	0.98	1.17	1.37
Government bodies	0.4	0.53	0.7
Developing automation	0.23	0.3	0.44
In-person security	0.83	0.95	1.1
Manufacturing, natural resources	0.33	0.4	0.5
Automotive industry	0.27	0.36	0.47
Health care suppliers	0.2	0.28	0.36
Retail, wholesale trade	0.3	0.36	0.44
Information technology	0.37	0.37	0.37
Transport and logistics	0.06	0.07	0.08
Total	**3.96**	**4.82**	**5.82**

[Source: zdnet.com].

up short and to trade it prior to any chances of getting hurt. Affiliations also additionally utilize the data made through sensors to create relevant structures, along with its stock chains, for more reasonable implications in light of the fact that it could lead to incredibly larger authentic data with respect to the current whereabouts.

The Industrial Internet of Things (IIoT), otherwise called fourth mechanical agitation or Industry 4.0, is specific nomenclature provided for the usage of IoT development for businesses. This idea can be compared to for the purchaser IoT contraptions in the home, yet for the present circumstance, the fact is to use a mix of sensors, far off associations, tremendous data, Artificial Intelligence, and assessment to measure and propel current cycles. At whatever point introduced over an entire store organization, rather than essentially particular associations, the impact could be altogether more noticeable inside the last possible second transport of materials and the organization in creation from beginning till end. Growing employee base efficiency or holding down expenses are two probable focuses, yet the IIoT may similarly make novel cash flows with associations; instead of essentially parting with a free thing, i.e., an engine, creators can in like manner sell it.

The IoT promises to make our current conditions, homes, workplaces, and vehicles smarter, more quantifiable, and chattier. Devices such as Google Home and Amazon Echo facilitate playing music, setting time tickers, or getting data. House security systems make it simpler to screen what's happening internally and externally or to see and talk with guests. Smart indoor regulators help by heating our homes while we're out, and smart lights can make it seem as if someone is home any time while one is out. Beyond the home, sensors can help us in seeing how clamorous or dirtied our current conditions may be. Auto-driven vehicles and mind-blowing metropolitan domains could change how we assemble and oversee

Trade-Offs and Vulnerabilities

public spaces. These degrees of progress could have gigantic ramifications concerning system security.

4.5 RISKS AND BREACHES OF IOT DEVICES

For each and every associated sensor accumulating data on everything that is done, the IoT is conceivably both a colossal protection and a security migraine. Consider a smart home: the suggestions are apt for making coffee (smart espresso machine has begun) and the effectiveness of tooth cleaning (because of a smart toothbrush), which radio broadcast one looks at (thanks to a smart speaker), which kind of food one consumes (by virtue of a smart stove or cooler), one's assessment (as a result of their smart toys), and who are our regular visitors, who comes near our homes (by prudence of a smart doorbell). Just as companies continue to get cash from selling the smart article in any case, their IoT procedure presumably joins selling in all probability a touch of that information, too.

Security concerns have troubled the Internet-of-Things (IoT) since the term was coined. Everyone from traders to huge business customers to buyers is worried that the latest IoT devices and systems are being subverted. The bug is considerably horrible compared to what is known, just as powerless IoT devices may be hacked and handled through goliath botnets, which subvert properly protected companies. But what are the most difficult issues and shortcomings to avoid when amassing, sending, or managing IoT systems? Moreover, what might we have the option to do to reduce these errors?

In this context, the Open Web Application Security Project (OWASP) is taken into consideration, as mentioned in generic OWASP documentation.

The OWASP Internet of Things Project is intended to help producers, engineers, and customers better comprehend the security issues related with the Internet of Things, and to empower clients in any setting to settle on better security choices when constructing, conveying, or surveying IoT innovations.

Vulnerabilities of the IoT include:

- A frail, guessable, or hardcoded secret phrase
- Uncertain network service
- Uncertain ecosystem interfaces
- Absence of secure update mechanism
- Utilization of insecure or outdated components
- Inadequate privacy protection
- Uncertain data transfer and storage
- Absence of device management
- Uncertain default configurations
- Absence of physical hardening

52 Cybersecurity

According to the OWASP IoT project, the 16 most dangerous IoT threats clustered based on security aspects are:

4.5.1 VERIFICATION

i. absence of objects to evade username count: attacker can gather legitimate account details via verification instruments.
ii. utilizing feeble passwords: this alludes to passwords utilized for a certain confirmation cycle opposing a part of the IoT administration, which could lead to anything but are difficult to figure out, on account of the low level of intricacy.
iii. absence in a record lockout object, after multiple bombed trials: this will permit a possible attacker to verify identity commonly without being hindered.
iv. absence of two-factor validation for basic capacities: this means that the confirmation measures engaged with the IoT administration don't consider multi-verification to get to basic capacities.

4.5.2 CRYPTOGRAPHY

i. decoded network administrations permitting listening to correspondence between parts of the IoT administrations are not protected, conceivably permitting alteration.
ii. coming up short in the execution of encryption systems: encryption strategies are ineffectively actualized, inappropriately designed, or not actualized appropriately.

4.5.3 MODIFYING TECHNIQUES

i. modifications shared without abstraction: refreshes in various design layers can be communicated throughout the organization instead of utilizing TLS or scrambling the records associated with the cycle.
ii. a distant change can be managed exclusive of security controls: no verification technique or secure controller is present to direct the modifications wider range devices.
iii. capacity area for refreshed documents is writable: a capacity area for update records has composed and perused authorizations to any client, permitting firmware change or dispersion.

4.5.4 PHYSICAL PERMIT

i. software embedded in hardware and information is separated permitting admittance in delicate data: firmware is removed or provides abundant data compared to normal scenarios.
ii. conceivable admittance for devices reassures on account of an absence of objects: it is conceivable to get the entire comfort permit via sequential interfaces, or absence of objects to try not to enter the single-client mode.

Trade-Offs and Vulnerabilities 53

iii. capacity source can truly be unprotected and is, at times, eliminated.
iv. absence of objects for keeping away from actual association with the device, causing control with respect to program running stream: this can be thought to change the running program to sidestep security objects or admittance in case of sensitive data.

4.5.5 SYSTEM CONTROL

i. absence of objects opposing Denial of Service: this can be seen as refusing assistance via the organization or through the device.
ii. absence of objects for maintaining a strategic distance from order infusion: it includes no counterarguments for avoiding intrusions, e.g., an order or SQL, that influences specific information overseen by the IoT administration.
iii. IoT administration contains uncertain outsider segments: applications or segments created by outsiders are obsolete or have security shortcomings.

A person could undoubtedly contend that the task of forestalling and responding to these weaknesses isn't minor. In the IoT setting, it involves a specific requirement to help specific framework overseers, which carries the burden of controlling several distinct occasions coming from various sources. All in all, the security investigation of the IoT environment is unpredictable, since it is presented with various sorts of attacks, utilizing diverse attack surfaces and misusing an assortment of weaknesses present in IoT frameworks. In any case, the IoT conditions produce a tremendous number of various occasions that should be utilized to expand the security level of the whole framework.

4.6 CLOUD COMPUTING

The concept of cloud processing has floated in our society since the mid-2000s, yet dealing with companies has been around for much longer: as early as the 1960s, when PC workplaces would permit the relationship to lease time on a focused device, instead of needing to get one themselves. The mentioned "time-sharing" associations were all around overwhelmed by the move to PCs, which soon made the ownership of a personal computer more reasonable, and thus the ascent of corporate data farms, where companies could store huge amounts of information. In any case, leasing acceptance to enrolling power has returned over and over: in the application master networks, the utility of framework outline through the end of the 1990s and mid-2000s. This was followed by conveyed handling, which truly took hold with the improvement of programming, as help and hyper-scale scattered enlisting suppliers, e.g., Amazon Web Services.

Distributed computing is considered to be the movement of on-demand enrolling organizations going from applications to limit and getting ready force which is ordinarily open throughout the Internet in a refund relation that is every bit while expenses grow on scenarios, Figure 4.2. Instead of asserting particular registering system otherwise worker ranches, affiliations are given permission to anything through programs to restrict within the reaches of a cloud expert center. One great result of

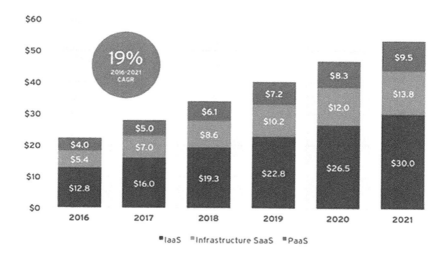

FIGURE 4.2 Cloud computing "as-a-Service" revenue ($bn).

[Source: 451 Research's market monitor, cloud computing, November 2017].

utilizing appropriated preparing associations is that affiliations can maintain a strategic distance from the immediate expense and erraticism of affirming and keeping up their own IT frameworks and rather pay only for what they use, when they use it. Therefore, suppliers of circulated preparing associations can profit by fundamental economies of scale by giving similar associations to a wide degree of clients.

Its associations cover a colossal degree of choices currently, from the fundamentals of cutoff, structures association, and managing power to customary language preparing and man-made awareness comparatively as standard office applications. Essentially, any idea that you ought to be truly near the PC equipment that you are using could now be replaced by the choice of going through the cloud. Dispersed calculation supports an enormous number of associations. That wires client associations such as Gmail or the cloud back-up of the photographs on your telephone any route to the associations which permit tremendous endeavors to have all their information and run their applications in the cloud. Netflix depends upon appropriated figuring associations to run its video electronic segment, and its other business structures likewise and has various affiliations. It is changing into the default elective for specific applications: programming shippers are legitimately offering their applications as associations over the web, rather than self-governing things, as they attempt to change to a subscription model. Regardless, there is a possible drawback to scattered handling, in that it can also present new expenses and new dangers for affiliations utilizing it.

Building up the foundation to help circulated figuring as of now addresses an excess of 33% of information technology spending all over, as indicated by research from IDC. Meanwhile, planning on conventional, in-house IT keeps sliding, as enrolling remaining positions waiting be done keeps moving to the cloud, regardless

Trade-Offs and Vulnerabilities

of whether that is public cloud associations offered by sellers or their private clouds. 451 studies hint that about 33% of enormous business IT spending will be on empowering and cloud benefits this year "demonstrating a making dependence on outer wellsprings of foundation, application, the heads and security associations." Gartner Research predicts that a significant amount of the general undertakings utilizing the cloud currently will have bet everything on it by 2021. As indicated by Gartner, in general, spending on cloud associations will reach $260bn this year, up from $219.6 billion. It's likewise happening faster than the researchers anticipated. Regardless, it's not away from a ton of that requesting is coming from affiliations that need to move to the cloud and what is being made by brokers who, as of now, offer cloud varieties of their things (as frequently as possible, since they race to stop offering one-off licenses, instead selling perhaps more profitable and visible cloud investments).

4.7 VULNERABILITIES WITH CLOUD OFFERINGS

Affiliations keep growing new applications or moving existing offerings to cloud-native associations. The public governments are making a cloud division a focal, basic part of the IT modernization structure. A connection that gets cloud moves and moreover picks cloud master affiliations and associations or applications without getting completely taught concerning the hazards included opens itself to a multitude of business, monetary, explicit, bona fide, and solid prospects. In this section, we chart 12 dangers and inadequacies that affiliations face while moving applications or information to the cloud.

NIST distinguishes the accompanying qualities and models for distributed computing:

- Essential features: done after getting a request for its own administration, expansive company permit, asset pooling, quick flexibility, and approximated administration
- Service model: Software as a Service (SaaS), Infrastructure as a Service (IaaS), and Platform as a Service (PaaS).
- Availability models: private, network, public, and crossbreed

4.7.1 LESSER CLARITY WITH CONTROL

When propelling a set of resources/tasks in a cloud, the alliances lose considerable permeability and authority over those resources. While utilizing outer cloud benefits, the commitment regarding a fragment of the plans and foundation moves to the supplier. The genuine shifting of responsibility relies on the cloud association model(s) utilized, prompting a change in stance for clients, contrasting with security inspection and logging. Your association needs to perform seeing and assessment of data about applications, associations, information, and clients, without utilizing network-oriented verification and logging, which is open for the physical storage of an IT scenario.

4.7.2 On-Demand Self-Service

Suppliers make it easy to orchestrate current businesses. The on-request self-association allocation highlights of the cloud draw in your alliance's workers to arrange extra associations from the supplier without IT assent. This display of utilizing programming in a connection that isn't maintained by the association's IT division is called shadow IT. Considering the lower expenses and ease of finishing PaaS and SaaS things, the likelihood of the unapproved utilization of cloud associations increases incrementally. Associations provisioned or utilized without informing IT present dangers to a connection. The utilization of unapproved cloud associations could accomplish an expansion in malware contaminations or information exfiltration, since your connection can't validate assets it doesn't recognize. The utilization of unapproved cloud benefits also reduces your connection's vulnerability and control of affiliation and information.

4.7.3 Worldwide Controlling APIs

Providers uncover many application programming interfaces (APIs) that clients use to oversee and interface with cloud associations (called the association plane). Affiliations utilize these APIs as a strategy to control, organize, and screen their resources and clients. These APIs can contain equivalent programming inadequacies as an API for a working structure, library, and so on. Unlike the central APIs for on-premises planning, supplier APIs are open, utilizing the Internet, exposing them much more generally to misuse. Dangerous actors search for weaknesses in association APIs. At whatever point they are found, these deficiencies can be the basis of fruitful attacks, and a connection's cloud resources can be undermined. Beginning there, attackers utilize association resources to execute further attacks against different clients of the supplier.

4.7.4 Multi-Tenant Feature

The unethical use of framework- and code-oriented shortcomings internal to a provider's establishment, levels, or applications that supports multi-residency incite errors and establish a division between tenants. This disappointment can be utilized by an attacker to gain entrance, starting with one association's asset then onto coming up next client's or connection's resources or information. Multi-residency develops the attack surface, inducing an all-encompassing possibility of information spillage if the division controls fail spectacularly. This attack can be refined by exploiting deficiencies in the supplier's applications, hypervisor, or gear, disturbing sensible division controls, or attacks on the supplier's association API. No reports of an attack dependent on clear partition frustration have been perceived; regardless, affirmation of thought mishandles have been addressed.

4.7.5 Information Removal

Threats related to data erasure continue to exist, considering that the purchaser decreases the perceivable standard inside where the relevant information gets dealt

Trade-Offs and Vulnerabilities 57

within the cloud along with specific lessened capacity for certifying a certain dropping of their information. The danger is emphasized in light of the fact that specific data becomes scattered over a number of groups putting away appliances internal to the supplier's foundation in a shared environment. Additionally, dropping structures may vary from supplier to supplier. Affiliations will without a doubt not be ready to avow that their information was safely erased and that remainders of the information are not accessible to attackers. This hazard increases incrementsally, as a client utilizes more supplier associations.

4.7.6 STOLEN USER DETAILS

On the off chance that an intruder gets to one of your client's cloud capacities, attackers push towards the supplier's associations to give extra assets (considering that the accreditations permitted consent to course of action), correspondingly as consideration on your connection's resources. The attacker could use scattered assets to revolve around your alliance's real clients, different affiliations utilizing a near supplier, or the supplier's chiefs. An intruder who gets to a supplier manager's cloud abilities may have the choice to utilize those accreditations to get to the clients' frameworks and information. The chief positions change between a supplier and a connection. The supplier's head advances toward the supplier affiliation, frameworks, and applications (subordinate upon the association) of the supplier's foundation. By and large, the supplier chief has managerial rights over more than one client and supports different associations.

4.7.7 SUPPLIER COMMITMENT

This changes in a bug where your association considers moving its resources/tasks initiating with one supplier then onto the accompanying. This association will likely find that the cost/exertion/plan time-critical of a specific move is altogether larger than from the beginning associated with an immediate outcome with factors, e.g., nonstandard information plans, nonstandard APIs, and dependence on one supplier's prohibitive instruments and unprecedented APIs. This issue increases incrementally in assistance models where the supplier acknowledges more conspicuous commitments. As a client utilizes more highlights, associations, or APIs, the prologue to a supplier's novel execution increases incrementally. Such executions require changes when a cutoff is moved to a substitute supplier. In the event that a chosen supplier withdraws its help because of a loss of business, it changes into a basic issue, since information can be lost or will not be ready to be moved to another supplier in an ideal way.

4.7.8 HIGHER COMPLEXITY

Embracing the cloud gives IT tasks a multifaceted nature. Managing, putting together, and running in the cloud may necessitate that the association's current IT staff get to know another model. IT staff should have the cutoff and authority level to oversee, join, and keep up the relocation of resources and information to the cloud, in spite of their present responsibilities regarding on-premises IT. Key associations and encryption associations become more unpredictable in the cloud. The associations, procedure, and devices accessible to log and screen cloud benefits normally change across suppliers, further developing the multifaceted nature. There may likewise be dangers/opportunities in cloud usage considering progression, approaches, and execution methodology, which add a multifaceted nature. This additional unpredictability prompts an all-encompassing potential for security openings in an affiliation's cloud and on-premises usage.

4.7.9 INSIDER ABUSE

Related faculty, e.g., employees, and bosses for the two affiliations and suppliers, who misuse their certified enlistment to the alliance's or supplier's affiliations, structures, and information are peculiarly orchestrated to cause harmed or exfiltrate data. The effect is no uncertainty more frightful while utilizing IaaS because of an insider's capacity to arrange assets or perform terrible exercises that require bad behavior scene assessment for revelation. These legitimate cutoff points might be inaccessible with cloud assets.

4.7.10 LOST INFORMATION

Information held through the cloud has the possibility of being lost for causes alternative to unsafe attacks. The unintentional destruction of information by the cloud master affiliation or a genuine upheaval, e.g., a fire or any tremor, can incite the permanent loss of client information. The burden of evading information difficulty doesn't fall just on the supplier's shoulders. If a client scrambles its information prior to moving it to the cloud, or loses the encryption key, the information will be lost. Furthermore, a deficient view of a supplier's scoring model may cause an information accident. Affiliations ought to think about information recovery and be ready for the chance of their supplier being gotten, changing help responsibilities, or failing. This danger increases incrementally as an association utilizes more supplier associations. Recovering information from a supplier might be less mind-boggling than recovering it at an office, considering the way that an SLA gives out accessibility/uptime rates. These rates ought to be investigated when your association picks a supplier.

4.7.11 PROVIDER SUPPLY CHAIN

On the off chance that your supplier reevaluates portions of its structure, tasks, or upkeep, these outsiders may not fulfill/keep up the necessities that the supplier is contracted to compel the connection to run. Your connection needs to assess how the supplier favors consistency and check whether the supplier streams its necessities down to the rejects. If the necessities are not being accumulated on the store association, then the danger to your connection increases incrementally. This risk increases as your alliance utilizes more supplier benefits and becomes reliant upon lone suppliers and their creation network systems.

Trade-Offs and Vulnerabilities 59

4.7.12 Inadequate Due Perseverance

Affiliations adopting the cloud every now and again perform insufficiently because of steadiness. The transformation information to the cloud instead of comprehending the full level of doing taking everything into account, the security attempts utilized by the supplier, and their commitment to give prosperity tries. They settle on choices to utilize cloud associations without thoroughly seeing how those associations ought to be guaranteed.

4.8 SECURE CLOUD COMPUTING TECHNIQUES

Based on the aforementioned vulnerabilities, this section covers ambient techniques that help create a secure cloud computing environment. We take a look at different aspects of infrastructure, application and data security, and middleware security that impacts the system, as shown in Figure 4.3.

4.8.1 Infrastructure Security

4.8.1.1 Physical Security

To secure your server farm, you should guarantee that you are confining admittance to the office to just approved people. Explicit advances incorporate access control estimates, e.g., access cards, every minute of every day, video reconnaissance checking, and an on-location security group, for a start. You should likewise have possibilities set up to forestall information risks brought about by catastrophic events, an on-location episode, loss of influence, and different dangers. within these possibilities, an information recovery plan is essential, alongside various redundancies.

4.8.1.2 Network Security

One should consolidate network checking, separating, and access control to disengage pernicious virtual machines, relieve the conveyed refusal of-administration (DDoS) attacks, and prevent dubious access/logins. In this regard, you should

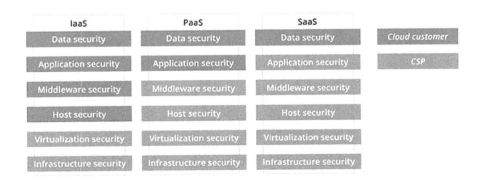

FIGURE 4.3 Types of cybersecurity for cloud computing.

[Source: AboutSSL, https://aboutssl.org/the-current-security-techniques-in-cloud-computing].

60 Cybersecurity

introduce firewalls, security entryways, and DDoS frameworks and pair those with an organization security group to screen and speedily react to episodes.

4.8.2 REMOTE SECURITY

The objective of remote security goes towards holding application programming interfaces (APIs) secure also as discrete tenants in your distant devices or compartments. It should likewise guarantee that virtual affiliation transmissions are secure. In addition, strategies with network security are supported by firewalls and restricted to DDoS frameworks. This is a chance also to look at the unpleasant pollution gateways and intrusion prevention systems, or IPS.

4.8.3 HOST SECURITY

The idea covered in this segment concerns guaranteeing what you're empowering on your workers, e.g., customer information, application information, APIs, and so on, from threatening code, obstruction, and weaknesses. A standard responses to this issue is to merge a secure framework, to perceive the attacker entering into the IPS and sandbox toxic code, to fix any update. Notwithstanding, security isn't bound only to frameworks; there's a preparing section including you to:

- Debilitate waiting or unutilized ports.
- Present, along with maintaining, access control.
- Dispense with unnecessary cycles and old cases.
- Notice logs and scenes.

4.8.4 SECURITY FOR MIDDLEWARES

4.8.4.1 Containers

For owners, the fundamental goal (considering that the establishment security is presently set up) is to restrict permission to the vaults. Furthermore, you should ensure that the compartments are viably orchestrated, to hinder slip-ups and shortcomings.

4.8.4.2 Application Programming Interfaces (APIs)

Validating APIs may consist of steps, cycles, and frameworks that are set up to accomplish the following aspects:

- Verify APIs to ensure only authentic API requests will gain access, and subsequently, rejecting the questionable ones.
- Notice APIs: one should design a structure set up for checking and verifying APIs, especially for essential API prosperity estimations, e.g., the bungle rate and deferrals.
- Programming interface log keeping: the ideal concept here is to check and verify what/when/who are conjuring the APIs, ensuring data gets sent securely, and check for questionable development.

4.8.4.3 Databases

To make sure about information bases, your beginning stage should be to direct information base reviews. Certain surveys are taken into consideration for verifying security dangers introduced in database systems, screen direct, along with setting up a scene noticing and prepared device.

4.8.4.4 Resource Management Platform

The purpose here for having a resource in specific control stages is for checking, verifying, perceiving, and responding to possible questionable and unapproved activity. It is started by executing the given platform within an internal security information and event management (SIEM) measure.

4.8.5 APPLICATION SYSTEM SECURITY

The expectation here is to make sure about what one is encouraging for stakeholders—e.g., customer, program information, APIs, data, etc.—via pernicious programs, interference, and shortcomings. General and quality responses in similar issues join an adversary for contaminating system resources to perceive harmful programs entered through IPS, to make the changes. Regardless, having security isn't confined to simply having structures; it also consists of taking care of:

- User Authentication
- Programming interface permit security
- Record management
- User role authentication

4.8.6 DATA SECURITY

Security has the aim of shielding data from delivery or damage. Additionally, the owner should guarantee that network transmissions are secure and that your information is protected from creation, burglary, changes, and other dangerous exercises. The fundamental development is to encode the information in its accumulating and transmission structures. Furthermore, you should guarantee that the information of each occupant is disengaged from the others'. As demonstrated beforehand, you could comparatively execute an information base review to see the expected dangers and block them early. The subsequent development is to guarantee that the information is essentially being sent safely. In general, for safety concerns, it passes SSL/TLS through the use of HTTPS. In all honesty, this is an administrative need in PaaS or SaaS-set up responsibilities consolidated regarding the web parcel, e.g., Stripe and Shopify. Post this, a following third process consolidates cycles, e.g., access control and others to guarantee that single confirmed people (to the excusal of all others) approach the information and that additionally if it's needed for their work. Finally, you should have assessments set up to recover that information in the event of a breach or other fiasco (i.e., calamity recovery).

4.9 SUMMARY

This chapter on trade-offs and vulnerabilities in IoT and secure cloud computing covers a host of topics. The reader gets an idea of works and research done on the subject, followed by an in-depth discussion of the cloud and IoT and their vulnerabilities. The topics covered are:

- Cloud computing and the IoT define the way we live, with the networking of devices and the ability to access all our needed data and information from anywhere and any device.
- There are vulnerabilities and risks associated with all such scenarios.
- There is a lot of research done in this field, and we cover some related research as part of this chapter.
- We took a look at a history of breaches involving some very high-profile names, highlighting the importance of cloud security.
- The subsequent section talks about the Internet-of-Things and how it has taken over the ease in connectivity.
- Based on the premises on IoT, we take a look at possible breaches that can affect a network of such devices.
- We then discuss the concept of cloud computing and provide a brief about the importance of the cloud in today's scenario.
- We look at 12 vulnerabilities of the cloud, of which some are unique to the cloud and some could happen for both on-premise and cloud systems, alike.
- The risks and vulnerabilities that can affect or have an impact on cloud systems are discussed in detail.
- Finally, we explore segments of the cloud that can be made secure through different tools and techniques in the buckets of infrastructure, virtualization, host and middleware, application systems, and data security.

REFERENCES

Bradford, C. (n.d.). "7 Most Infamous Cloud Security Breaches." Available: https://blog.storagecraft.com/7-infamous-cloud-security-breaches/ (Last Accessed: 09.12.2020)

Chen, C., and Wang, C. (2019). "*Constructing of Vulnerability Prevention Secure Model for the Cloud Computing,*" *2019 6th International Conference on Systems and Informatics (ICSAI),* Shanghai, China, 2019, pp. 694–698. DOI: 10.1109/ICSAI48974.2019.9010146

De, S. (2015). "Addressing General Data Protection Regulations for Enterprise Applications Using Blockchain Technology," *International Journal of Applied Research on Information Technology and Computing,* vol.10, no. 2, pp.67–75. ISSN: 0975-8070, DOI: 10.5958/0975-8089.2019.00009.5

De, S. (2020a, September). "Security Threat Analysis and Prevention towards Attack Strategies." In *Gautam Kumar, Dinesh Kumar Saini, and Nguyen Ha Huy Cuong (Eds.), Cyber Defence Mechanism,* Boca Raton: CRC Press, DOI: 10.1201/9780367816438-1

De, S. (2020b). "*A Novel Perspective to Threat Modelling using Design Thinking and Agile Principles,*" *2020 Sixth International Conference on Parallel, Distributed and Grid Computing (PDGC),* Waknaghat, Solan, India: IEEE, pp. 31–35. DOI: 10.1109/ PDGC50313.2020.9315844 henricoDolfing. (n.d.) "Cloud Computing Threats, Vulnerabilities and Risks." Available: https://www.henricodolfing.com/2019/03/cloud-computing-threats-vulnerabilities.html (Last Accessed: 12.12.2020)

López, D. D., Uribe, M. B., Cely, C. S., Torres, A. V., Guataquira, N. M., Castro, S. M., Nespoli, P., and Mármol, F. G. (2018). "Shielding IoT against Cyber-Attacks: An Event-Based Approach Using SIEM," *Wireless Communications and Mobile Computing*, vol. 2018, Article ID 3029638, DOI: 10.1155/2018/3029638

Mishra, N. and Singh, R. K. (2019). "*Taxonomy & Analysis of Cloud Computing Vulnerabilities through Attack Vector, CVSS and Complexity Parameter*," *2019 International Conference on Issues and Challenges in Intelligent Computing Techniques (ICICT)*, Ghaziabad, India, pp. 1–8. DOI: 10.1109/ICICT46931.2019.8977667

Morrow, T. (n.d.). "12 Risks, Threats, & Vulnerabilities in Moving to the Cloud." Available: https://insights.sei.cmu.edu/sei_blog/2018/03/12-risks-threats-vulnerabilities-in-moving-to-the-cloud.html (Last Accessed: 12.12.2020)

Moustafa, N., Turnbull, B., and Choo, K. R. (2019, June). "An Ensemble Intrusion Detection Technique Based on Proposed Statistical Flow Features for Protecting Network Traffic of Internet of Things," *IEEE Internet of Things Journal*, vol. 6, no. 3, pp. 4815–4830. DOI: 10.1109/JIOT.2018.2871719

OWASP Internet of Things, (2021). Available: https://www.owasp.org/index.php/OWASP_Internet_of_Things_Project (Last Accessed: 25.01.2021)

Paikrao, R. L. and Patil, V. H. (2018). "*Security as a Service Model for Virtualization Vulnerabilities in Cloud Computing*," *2018 International Conference on Advances in Communication and Computing Technology (ICACCT)*, Sangamner: IEEE, pp. 559–562. DOI: 10.1109/ICACCT.2018.8529573

Pal, S. K. and De, S. (2015a April 1). "A Ciphering Algorithm on Squaring and Sine Function using Sequence of Random Numbers," *International Journal of Applied Research on Information Technology and Computing*, vol. 6, no. 1, pp. 50–62. Print ISSN: 2249-3212. Online ISSN: 0975-8089 DOI: 10.5958/0975-8089.2015.00008.1

Pal, S. K. and De, S. (2015b). "An Encryption Technique Based upon Encoded Multiplier with Controlled Generation of Random Numbers," *IJCNIS*, vol.7, no. 10, pp. 50–57. DOI: 10.5815/ijcnis.2015.10.06

Panjwani, M. and De, S. (2020). "*Study of Cloud Security in Hyper-Scalers*," *2020 7th International Conference on Computing for Sustainable Global Development (INDIACom)*, New Delhi, India: IEEE, pp. 29–34. DOI: 10.23919/INDIACom49435.2020.9083727

Paul, F. (n.d.). "Top 10 IoT Vulnerabilities." Available: https://www.networkworld.com/article/3332032/top-10-iot-vulnerabilities.html (Last Accessed: 12.12.2020)

Ranger, S. (n.d.-a). "What Is the IoT? Everything You Need to Know about the Internet of Things Right Now." Available: https://www.zdnet.com/article/what-is-the-internet-of-things-everything-you-need-to-know-about-the-iot-right-now/ (Last Accessed: 09.12.2020)

Ranger, S. (n.d.-b). "What Is Cloud Computing? Everything You Need to Know about the Cloud Explained." Available: https://www.zdnet.com/article/what-is-cloud-computing-everything-you-need-to-know-about-the-cloud/ (Last Accessed: 12.12.2020)

Sahmim, S. and Gharsellaoui, H. (2017). "Privacy and Security in Internet-based Computing: Cloud Computing, Internet of Things, Cloud of Things: A Review," *Procedia Computer Science*, vol. 112, pp. 1516–1522, ISSN 1877-0509, DOI: 10.1016/j.procs.2017.08.050

Stevens, J. (2020). "What Are the Current Security Techniques in Cloud Computing?" Available: https://aboutssl.org/the-current-security-techniques-in-cloud-computing (Last Accessed: 12.12.2020)

Stocker, E. and Walloschek, T. (n.d.). "Understanding Cloud Computing Vulnerabilities." Available: https://www.infoq.com/articles/ieee-cloud-computing-vulnerabilities/ (Last Accessed: 12.12.2020)

Wang, H., Chen, Z., Zhao, J., Di, X., and Liu, D. (2018). "A Vulnerability Assessment Method in Industrial Internet of Things Based on Attack Graph and Maximum Flow," *IEEE Access*, vol. 6, pp. 8599–8609. DOI: 10.1109/ACCESS.2018.2805690

Yin, D., Zhang, L., and Yang, K. (2018). "A DDoS Attack Detection and Mitigation with Software-Defined Internet of Things Framework," *IEEE Access*, vol. 6, pp. 24694–24705. DOI: 10.1109/ACCESS.2018.283128.

5 Location and Availability Protections in Smart Mobility

Praveen Gupta and Reena Sharma
Poornima Institute of Engineering and Technology, India

CONTENTS

5.1 Introduction 66
    5.1.1 Key Principles of Smart Mobility 66
    5.1.2 How Does Smart Mobility Connect to Smart City 67
    5.1.3 Smart Mobility and the Role of Data..68
5.2 Definition of Smart City...68
    5.2.1 Smart Cities: A Futuristic Vision 68
    5.2.2 Need for Smart City 69
    5.2.3 Successful Smart City 69
5.3 Technology for Smart Mobility..70
    5.3.1 Technical Characteristics of Smart City Services...............................70
    5.3.2 IoT Device Characteristics..70
    5.3.3 IoT Technology 71
    5.3.4 IoT for Smart City...72
    5.3.5 Examples of IoT Mobility Solutions...72
    5.3.6 Wireless Technology for Smart Cities 72
    5.3.7 Impact of Artificial Intelligence 73
5.4 Transportation and the Traffic Problem 73
    5.4.1 Services Using Mobility 73
5.5 Regulatory Characteristics for Smart City 74
    5.5.1 Government Policy and Legal Issues...74
    5.5.2 Common Legal Framework 74
    5.5.3 Government Policy across the Developed Nations 76
    5.5.4 Regulatory Characteristics of Smart City Services.............................76
5.6 Ecosystem of Smart City Services 77
5.7 Mobility as a Service (MaaS)...77
5.8 Security and Privacy of Data..77
    5.8.1 Four Core Security Objectives 78
    5.8.2 Five Types of Privacy 78
    5.8.3 Building Blocks for Privacy Protection 79

DOI: 10.1201/9781003145042-5

5.8.4	Privacy Techniques 80
	5.8.4.1 Process-Oriented Privacy Protection 80
	5.8.4.2 Data-Oriented Privacy Protection 81
5.9	Conclusion 81
References	 81

5.1 INTRODUCTION

The term "mobility" has been used for centuries; but, with the availability of new technologies, the meaning of mobility and the facilities associated with it has changed. Smart mobility is a relatively new term that entered the picture in the last decade. This term has been variously defined. Smart mobility is a new and revolutionary process of thinking about how we get around, one that is cleaner, safer, and comparatively more efficient. The concept of smart mobility is gaining ground inside and outside the scientific community. Around the world, the movement of people and things is called mobility. Smart mobility is a concept where, with various past and real-time data, and with the help of information and communication technologies, travel time is optimized, resulting in reductions of space usage, road congestion, road accidents, and emissions of harmful gases (Davor Brčić, 2018).

Mobility has many dimensions: intellectual, social, professional, and spatial. It also provides countless opportunities, constraints, and freedoms that shape modern society over time and across space. Smart mobility is known by the accessibility of information and communication infrastructure, through the development of innovative, sustainable, and safe transport. The smart environment is measured by the attractiveness of the natural environment, levels of pollution, resource management methods, and environmental protection activities. In a nutshell, it can be understood as travelling with the help of software and electronic devices, many electronic devices, which provide assistance.

5.1.1 KEY PRINCIPLES OF SMART MOBILITY

Smart mobility is based on the following key principles.

Principal	Description
Efficiency	The trip gets the traveler to their destination with minimal disruption and in as little time as possible.
Flexibility	Multiple modes of transportation allow travelers to choose which work best for a given situation.
Integration	The full route is planned door-to-door, regardless of which modes of transportation are used.
Safety	Fatalities and injuries are drastically reduced.
Clean technology	Transportation moves away from pollution-causing vehicles to following emission norms.

To provide a better quality of life to everyone, smart mobility includes two more aspects: accessibility and social benefit, meaning that it should be affordable to everyone.

5.1.2 How Does Smart Mobility Connect to Smart City

A smart city is a structure predominantly composed of information and communication technologies (ICT), to develop, deploy, and promote sustainable development practices to address the challenges of growing cities. The main part of this ICT structure is really an intelligent network of connected objects and machines that transmit data using wireless technology and the cloud. Cloud-based IoT applications receive, analyze, and manage data properly in real time, to help municipalities, enterprises, and citizens make better decisions that enhance the quality of life. Six thematic areas must be present and addressed in any smart city proposal:

- Smart people
- Smart mobility
- Smart economy
- Smart government
- Smart environment
- Smart living

These six areas are proposed in the model by Boyd Cohen, known as "the Cohen Wheel" (Luque-Vega et al., 2020). Smart people are characterized by their level of qualification, lifelong learning, and social participation in public life. Smart living is measured based on the existing cultural facilities, living conditions, health, safety, housing, educational facilities, tourist attractiveness, and social cohesion (Winkowska et al., 2019).

As explained earlier, smart mobility means a lot of communication devices. The newer term entering the picture is smart city, which means a city that provides the basic infrastructures to implement smart mobility. By using communication technologies and information, software, and business models, and to increase the operational efficiency of shared information, we can improve the quality of life for the residents of a smart city. One important pillar of the smart city concept is smart mobility, which is based on optimizing the transportation sector in urban areas. With smart mobility, more information is available and easy to receive, and life looks easier; but the current challenges must be addressed and appropriate plans developed to resolve them. Future challenges include:

- Vehicle traffic
- Public transportation
- Parking
- Freight
- Pedestrian traffic
- Emergencies

The majority of smart mobility systems are based on devices and data. The devices perform the following tasks:

- Data collection
- Data analysis
- Information transmission

5.1.3 Smart Mobility and the Role of Data

The availability of mobile devices with sensing capabilities has given rise to a number of platforms for mobile crowd sensing (MCS). According to Bellavista et al. (2015), MCS is commonly referred to as a paradigm for the distributed gathering of heterogeneous sensor data from pocket devices used by crowds. With the MCS solution, smart city managers can enable the monitoring of areas that are still not covered by fixed monitoring infrastructures. MCS allows gathering not only raw and locally processed sensor data but constructive comments and suggestions by residents.

5.2 DEFINITION OF SMART CITY

Because the smart city is based on fast-changing technology, there is no formal and widely accepted definition of a smart city. The ultimate goal is a better use of public resources, the improvement of the quality of services offered, and the reduction of the operational costs of public administration (Luque-Vega et al., 2020). A smart and sustainable city aims to become scalable, reliable, accessible, adaptable, and resilient (Luque-Vega et al., 2020).

The majority of the population lives in cities. The world is at an unprecedented level of urbanization. As per current statistics, the top 25 cities of the world today account for half of the world's wealth. Likewise, about 10 percent of the world population lives in the top 30 metro cities. People like to live in cities for many reasons, which may include available, convenient resources, good infrastructure, and the ease of doing business and working. Traditional city systems rely upon delivering resources. So, it is important to forecast future needs and develop systems for upcoming requirements. A smart city is perceived to be a harmonic modulation of history, with the availability of heritages, aesthetics, architecture, good information communication technology (ICT), fast delivery services, and improved accessibility. A smart city is a framework to optimize a resident's quality of life by leveraging technology and integrating functions such as intelligent transportation, data management, public safety, and security. Smart city deployments involve multiple features and state-of-the-art technologies and consist of diverse ecosystems from technology providers. Devices such as servers, sensors, gateways, and communication infrastructure collectively bringing to life the concept of the Internet of Things (IoT), a critical component in shaping the cities of the future.

5.2.1 Smart Cities: A Futuristic Vision

The term smart city has been in use since the 1990s. The term is defined differently by different authors. But a common element of these definitions is using information communication technology (ICT) to engage citizens, to deliver city services, and to enhance urban systems. A smart city is expected to be the key to merging a sustainable future with continued economic growth and job creation, to add a new identity and unique value to the lifestyle. According to Sarkar (2020), smart cities use digital technology to make urban systems more efficient, cost-effective, and environmentally

Location and Availability Protections 69

sustainable. Sensors embedded in buildings and infrastructure networks can help cities incorporate renewable energy, such as solar power, or save energy, for example, by turning on streetlights only when a road is in use. Sensors, smart cards, and digital cameras feed real-time data into integrated management systems; better data and analytic technologies can inform decision-making systems and improve management.

5.2.2 NEED FOR SMART CITY

As mentioned earlier, people are moving towards urban cities for a better lifestyle, so in order to handle the increased demand and workload, it is the requirement of our time to be prepared for technological challenges, so that cities can provide facilities and function smoothly. One need in the smart cities is to manage the large number of vehicles purchased by the public. Since cities are large, people need vehicles for local transportation. This large number of vehicles has created problems such as limited parking spaces in shopping centers and malls, speeding, not following traffic rules, and disobeying traffic signals. So, traffic is one area where a lot of work can be done, under the banner of the smart city. Another area is a mass transportation system for local travelers, such as office workers. A large transportation system needs to be developed for people, with metro trains, mono trains, etc.

Navigation is also one area where the smart city can help to find out the locations in better ways. Since a great deal of public and traffic is on the move, even in large cities, roads may be full of traffic during peak hours. Smart city technology could suggest alternative routes, minimizing traffic congestion. Smart technology could address many other urban needs. Mobile devices are the backbone of interacting with smart city network infrastructures, but they also present challenges for the security and privacy of users, where sensitive data could be vulnerable to attack by third parties (Ismagilova et al., 2020).

5.2.3 SUCCESSFUL SMART CITY

Cities implementing smart city projects include:

Amsterdam	Barcelona	Columbus, Ohio	Copenhagen
Dubai	Dublin	London	Madrid
Milan	Singapore		

In the aforementioned cities, the transformation to smart technologies is a work in progress and has not yet been fully completed anywhere. Four essential elements are necessary for smart cities:

1. Pervasive wireless connectivity
2. Open data
3. Security
4. Flexible monetization schemes

5.3 TECHNOLOGY FOR SMART MOBILITY

A vast number of different IoT services is used in different fields, from industrial settings to the government sector, and from people-centric to technology-oriented services. A classification of services according to their characteristics follows:

5.3.1 TECHNICAL CHARACTERISTICS OF SMART CITY SERVICES

The basic characteristics related to technical and other QoS parameters that are common to smart city services include:

1. Number of IoT devices: The number of devices that generate sensor data.
2. Number of end users: The number of users that use a service.
3. Data volume: The entire volume of generated data per service, which in turn generates traffic to servers hosting IoT platforms.
4. Platform openness and interoperability: IoT platforms should expose simple and unified interfaces, so that various innovative services and applications may integrate IoT devices, and corresponding data is managed by different IoT platforms.
5. Time sensitivity: Some of the services are sensitive to latency, like e-health services, where it is important to react immediately to specific events.
6. Location-based service: If a service depends on location, then it is necessary to foresee such a possibility and identify whether outdoor or indoor positioning is needed. For example, in the case of smart parking, the precise outdoor location of a car is necessary to identify an appropriate parking lot.
7. Scalability: Most cities are starting with a smaller city area, and thus it is necessary to design the software architecture for future expansions, in terms of increasing numbers of devices and users.
8. Billing: Service may be free of charge or, in some scenarios, end-users would pay for a service on a subscription basis.

5.3.2 IoT DEVICE CHARACTERISTICS

1. **Communication mode**: While designing a solution for city, a city planner should take into account the required communication infrastructure to ensure the connectivity of IoT devices to the Internet, the geographical characteristics of an area, the communication protocols available, and the services available. Adequate wired or wireless communication protocols should be used.
2. **Computing capability**: While installing a computing facility, data size must be properly estimated, and growth in the number of users and services must be considered. In the case where a device has sufficient computing capability, some operations and simple algorithms can be performed on it, while reducing communication with the server or platform. However, in addition to an increased price, such a device will also consume more power for computing. So, planners should carefully evaluate whether, for a specific service, additional computing

Location and Availability Protections 71

capability should be used on a device or whether the computation will be carried out within the cloud or at the network edge.

3. **Power consumption**: This is one of the most important characteristics of an IoT device. Devices are mostly battery powered. Due to the better efficiency of devices, and their low power mode of operations, as well as improved battery capacities, device-life has greatly improved. However, some devices still have to be connected to main power, due to their consumption characteristics. We are also seeing an improvement in energy-harvesting techniques, combined with low-power devices.

4. **Power source**: Power can be supplied to devices from a battery or main power supply, so a study of the availability of the power supply in a city must be made, and all mobile devices must have power supplied from batteries, while and fixed devices must have power backup.

5. **Location**: Location is identified using GPS technology, so all devices deployed for this purpose must be enabled with the GPS chipset. This is also dependent on some of the vendors, so those characteristics must be included.

6. **One-way/two-way communication**: Detailed description is needed, when the data is transmitted one-way and in which situation/circumstances data uses two-way transmission.

7. **Bandwidth**: Bandwidth decides the rate of data transfer. Planners should consider what the availability of bandwidth is to the service provider and what is available to the actual users. Since bandwidth changes rapidly, the complete system should be able accept those changes.

8. **Loss of data**: Some services may be highly sensitive to data loss, and thus, special mechanisms should be in place, e.g., retransmission, to protect against it. Some services might function correctly and be designed to be resilient to data loss.

9. **Delay**: Some services are sensitive to a delay in information delivery. For instance, real-time applications, such as smart parking, are highly sensitive to network delay.

5.3.3 IoT Technology

The IoT is a global infrastructure for the information society, enabling advanced services by interconnecting (physical and virtual) things based on existing and evolving interoperable information and communication technologies. IoT devices are equipped with communication devices, sensor network connection, and actuators, to work to achieve common objectives. but conceptually, it is similar to sensor networks. The IoT is different from other systems only in the sensing and communication capabilities, which are not the main feature of the object, but an extension to improve operation or provide additional services. Popular examples include smart meters, the smart fridge, and smart air conditioning systems. Evolutions in ICT and information sharing technology are the drivers of the scope and scale of today's smart city. This rapid evolution is revolutionizing smart city construction, with the dawn of the IoT.

5.3.4 IoT for Smart City

IoT technology and secure wireless connectivity are transforming such traditional components of city life as streetlights into next-generation intelligent lighting platforms with expanded capabilities. The scope of this includes integrating solar power and connecting to a cloud-based centralized control system that is connected to other devices in the ecosystem. High-power embedded LEDs alert the system about traffic issues, provide severe weather warnings, and provide a heads-up when environmental crises such as fires arise. Solar panels can provide energy to many devices, such as like streetlights, and also detect available free parking spaces and EV charging docks, and alert drivers where to find an open parking spot via a mobile app.

5.3.5 Examples of IoT Mobility Solutions

A smart city works with the cooperation and participation of three pillars of a city: industry, residents, and the government. Examples of objects that can fall into the scope of the IoT include connected security systems, thermostats, cars, electronic appliances, lights in household and commercial environments, alarm clocks, speaker system, vending machines, and many more. Businesses can leverage IoT applications to automate safety tasks (for example, to notify authorities when a fire extinguisher in the building is blocked).

5.3.6 Wireless Technology for Smart Cities

Given the access of mobile devices and smart applications, providing efficient mobility infrastructures and effectively managing mobility issues in future networks become more and more challenging. Long-term evaluation (LTE) 4G networks and the future 5G networks require smart mobility management schemes, to handle the mobility of even millions of mobile terminals. Mobility infrastructure and mobility management are also important in the new emerging computing paradigms, such as mobile cloud computing and fog computing. In recent years, along with the rapid development of location-based services (LBSs), accurate positioning techniques are required in many fields, including travel guidance, mobile advertising, and urban computing. In localization methods, a single type of signal feature as location fingerprint, which may not well deal with the instability of signal features caused by mobility. Compared to a single type of signal, we need to combine multiple types of signal features, to enhance the robustness and reliability of positioning and support more flexible mobility management. We classify mobility management schemes in WSNs into four categories:

1. Uncontrollable mobility (UCM)
2. Path-restricted mobility (PRM)
3. Location-restricted mobility (LRR)
4. Unrestricted mobility (URM)

Location and Availability Protections

5.3.7 Impact of Artificial Intelligence

Artificial intelligence (AI) is the basis on which human intelligence is imitated. For this purpose, algorithms are created, applied, and integrated into a dynamic computing environment. AI is an attempt to make computers think and behave humanly. As defined in 1956 by John McCarthy, AI is the science and engineering to create intelligent machines, especially intelligent computer programs, which is described as a field of study that gives computers the ability to learn without being explicitly programmed. Artificial intelligence enables a multitude of intelligent applications that make mobility safer, more comfortable, more efficient, and more resource-efficient. AI and machine learning technologies will greatly impact the automotive and mobility sectors, as they introduce new products and business models, rather than just improving productivity. AI technology:

1. Optimizes the control of traffic, systems, and vehicles
2. Improves the accuracy of traffic forecasts and weather forecasts
3. Increases the efficiency of logistics processes
4. Prevents accidents

Examples of AI in the Mobility Sector:

1. Reduction of occupational accidents
2. Public Transport on Demand
3. Predictive maintenance algorithms
4. Smart logistics

5.4 TRANSPORTATION AND THE TRAFFIC PROBLEM

Analyses of transportation issues are often constrained by domain-dependent data sources. Recent emerging technologies toward a connected vehicle-infrastructure-pedestrian (VIP) environment and using big data are making this task (collect, store, analyze, and use multisource data) easier. A connected VIP environment also makes the system more flexible, so that real-time management and control measures can be implemented to enhance system performance. The interchange of information occurs in four ways:

1. Vehicle-to-vehicle (V2V)
2. Vehicle-to-infrastructure (V2I)
3. Pedestrian-to-infrastructure (P2I)
4. Vehicle-to-pedestrian (V2P)

5.4.1 Services Using Mobility

Using mobile phone Wi-Fi signals, smart cameras, and open-source data, city planners and transportation managers can assess the level of crowding and, when necessary, reroute pedestrians. With smart phone penetration escalating each year, it is

74 Cybersecurity

plausible to use this to give transporters the power to anticipate loads and manage capacity, while empowering commuters to be certain of their position on a particular mode of transport, reducing the time spent in queues and sparing parties from exposure to risks such as the coronavirus, due to overcrowding.

5.5 REGULATORY CHARACTERISTICS FOR SMART CITY

The interest in improving mobility has increased, and governments and other institutions are trying to produce laws and develop long-term action plans.

5.5.1 GOVERNMENT POLICY AND LEGAL ISSUES

In smart city projects, sometimes, laws contribute to increasing and/or maintaining the smart mobility level. Such policies are part of the strategic plans proposed by institutions from America, Australia, and the European Union, and were retrieved in the same form as they were presented in the documents from the source column, without modification. Smart city project-related policy across the globe is seen in the following table.

5.5.2 COMMON LEGAL FRAMEWORK

A common legal framework is an operational procedure defined in the proposed system architecture. To better monitor the implementation of this framework, it is necessary to apply the decomposability feature of smart mobility governance. A common framework architecture, proposed by and presented in Docherty et al. (2018), divides governance into three categories, based on the decision-process level: 1. National, 2. Regional, and 3. Local government. The highest level of this layered architecture type shall adopt a national strategy, based on the country's priorities, economic development, current level of concept implementation, etc.

This strategy will be adapted to each region's needs by prioritizing the most important actions, taking into account the budget allocated. The lower level consists of the local government, which shall monitor the current status of smart governance

TABLE 5.1
Smart mobility policies and regulations

Region/Country	Agenda for smart city	Short description of the agenda
America	Sustainability	Technological development
	Low-carbon emissions	Control emissions
Australia	Autonomous driving	Policy and rules for autonomously driven cars
	Safety	Driving and safety rules
European Union	Autonomous driving	Navigation and road safety of the autonomously driven vehicles
	Charging infrastructure	Regulations for the charging infrastructure
	Data and publication	Implementation of open-access publication and data

Location and Availability Protections

focused on smart mobility challenges and adapt the regional strategy to local necessities. Moreover, local administrations will define:

1. Smart mobility business models, which will describe the city development purpose.
2. A procurement plan for the materials and necessary services to reach the defined goals.
3. The funds allocation for each decided measure, based on local budget and funds received from the upper layers.

The architecture for a smart mobility system is based on a common legal framework. To ensure the sustainability of the system, it is required to have a bottom-up feedback cycle between the layers proposed in this architecture. This will help introduce changes to the national strategy, according to the lower layers' feedback. Besides the changes in potential actions, the national strategy can tailor the allocation of funds based on the impact of priorities defined by local or regional governance, on the national economy, touristic needs, CO_2 emissions reduction strategies, or natural possible hazards exposure. Table 5.2 describes the role of government-level work and policy.

Deciding on a common worldwide framework for smart mobility implementation policies and regulations has advantages and disadvantages.

As advantages, we can mention:

1. A standardized legal framework for smart mobility concept implementation and evaluation
2. A simplified process for an objective comparison between different cities around the world, from a mobility perspective

TABLE 5.2
Government policy levels for a smart city

Government level	Policy and regulation classes
Local/Municipal corporation	Parking system
	Local transportation system
	Filling stations
	Traffic system
	Special areas
Regional/State Government	Traffic management centers
	Electric charging stations
	Fast connections between main cities
National/Country level	Vendor specifications
	Infrastructure specifications
	Autonomous driving
	Low-carbon emission
	Investment priorities
	Service provider regulation
	National regulatory body formation

76 Cybersecurity

3. Potential actions compliant with a common defined framework that can be decided by local administrations, to improve mobility in the city, reduce traffic congestion, improve residents' quality of life, etc.

Creating a common legal framework also involves some drawbacks. The greatest drawback could be the reduced number of policies and regulations that can be included in a common framework. This is due to policies and regulations for which implementations could be influenced by specific urban architecture, geographical position, natural hazards exposure, economical area development level, etc. However, this disadvantage can be seen also as an advantage, because the common framework offers a minimum set of laws that ensure the minimum requirements for having a smart mobility system. Each city could customize other possible actions, according to their needs and taking into account possible specific constraints.

5.5.3 Government Policy across the Developed Nations

Governments should feel the requirement and need for smart cities and act accordingly, while making privacy laws and regulations. Policies are under various stages of formulation in different countries. Many European countries have worked on them. Private partnerships will be more common in the near future. It is expected that many companies will participate in business. As companies enter into this business, governments need to formulate rules and regulations.

5.5.4 Regulatory Characteristics of Smart City Services

Regulatory characteristics are related to legal acts and laws relevant to smart city services. The European Union has formulated policies and guidelines, which can provide direction to other nations, also. Some of the characteristics for regulators can be summarized as follows.

1. **Lawful interception**: In many countries, national legislation requires the possibility of data traffic interception, which thus also applies to IoT data.
2. **Service dependability**: The ability to avoid existing service failures and the addition of new services.
3. **Personal data protection**: Personal data theft is very common these days, and regulators should make regulations to control data and its theft from service providers.
4. **Security**: Security of services, financial data, and other data is required.
5. **Operator switch**: Systems should be able to provide users the facility to change the service providers. Complete systems should provide multiple services and the facility to users to change service providers.
6. **Roaming**: Users can move from one service provider area to another service provider area, and in that situation, services could stop, due to having another service provider in that region. So, there should either be a national service provider or an ability to automatically switch to another service provider.
7. **Interoperability and open access to data and services**: Data common to users must be in open access to service providers. It is usable in the case of a switch in operators.

Location and Availability Protections 77

5.6 ECOSYSTEM OF SMART CITY SERVICES

An ecosystem involves the complete sustainability of the system, with the smooth functioning of all technical devices. The implementation of all government policy and regulatory bodies should be able to work smoothly. People should be able to use the system and receive benefits from it. The challenges include:

1. Mixture of technologies.
2. Autonomous driving.
3. Low emissions vehicles and alternative fuels.
4. Intelligent parking systems.
5. Smart infrastructure.
6. Vehicle-number increasing.
7. Citizen focus.
8. Integrated planning.
9. Business models, procurement, and funding.
10. Mobility as a service.
11. Supplying alternatives mean of transport.

5.7 MOBILITY AS A SERVICE (MAAS)

In every field of transport-related applications, Mobility as a Service (MaaS) has attracted attention. Driven by the Internet of Vehicles (IoV), vehicles gradually become mobile living spaces to satisfy residents' different demands, where the media functionalities, scenario scopes, and user engagements can be all expanded dramatically. MaaS provides ticketless and integrated transportation solutions, combining private cars, public transportation, shared travel, and walking. In order to meet user expectations, MaaS should be user-friendly, and service providers should understand user preferences. Depending on the digital platform, MaaS service aggregates numerous mobile applications, including the functionalities of planning, trip creation and management, ordering and payment, and real-time trip information. The first commercialized solution with an app operated by MaaS Global has been providing mixed transport and travel packages since 2016.

The plan selected is integrated with different transport modes, e.g., car-sharing, car rental, metro, rail, bus, bike-sharing, and taxi via the subscription of on-demand instant purchase, and an affordable monthly package can be paid by the user's account. MaaS combines the business model with ICT, to realize technology-enabled mobile service management. The MaaS solution is technology-sensitive and will evolve along with the development of more sophisticated wireless and intelligent technologies.

5.8 SECURITY AND PRIVACY OF DATA

This section focuses on security and related issues, to be more precise, what we want from service providers and regulatory services. Many entities participate in a smart city project, for example, governments, enterprises, software providers, device manufacturers, energy providers, and network service providers. They all integrate solutions and have to abide by four core security objectives.

5.8.1 Four Core Security Objectives

The core objectives of data security can be classified as:

1. Availability
2. Integrity
3. Confidentiality
4. Accountability

Availability: The project will use real-time, reliable data. Such projects cannot run on the basis of stored data. Data is being generated by users all the time, and it must be available to all parties involved in these services of the smart city project. Security solutions must avoid having negative effects on availability.

Integrity: Real data is the key of all services, so all devices must generate reliable data and have same technical standard to measure, collect and store data. Data should be nonmodifiable at all times.

Confidentiality: Data collected will have associated data of the individual, which is also an integral part of the information, otherwise, the information is not meaningful. Some of the data collected, stored, and analyzed will include sensitive details about consumers themselves. Steps must be taken to prevent the unauthorized disclosure of sensitive information.

Accountability: The users of a system must be responsible for any type of action taken. Their interactions with technique-sensitive systems should be logged and associated with a specific user. These logs should be difficult to forge and have reliable integrity protection.

To achieve these security core objectives, strong authentication and ID management solutions need to be integrated into the ecosystem, to ensure that data is shared only with authorized users.

These solutions also protect backend systems from intrusion and hacking.

5.8.2 Five Types of Privacy

Data privacy can be classified into five types:

1. Privacy of location
2. Privacy of state of body and mind
3. Privacy of social life
4. Privacy of behavior and action
5. Privacy of media

Privacy is considered a human right in every city. This is a very typical task, due to hacking technology where large amounts of private data are collected, processed, and stored. According to Eckhoff and Wagner (2017), five types of privacy have emerged:

1. **Privacy of location**. Location information is not data regarding location itself. It is data about when and for how long the location was visited. Privacy of location is based on spatial-temporal information. A violation of location privacy could give incorrect data about a person's home or workplace. A person's social life also depends on location information.
2. **Privacy of state of body and mind**: A person's bodily characteristics, such as biometrics, their health, mental states, emotions, opinions, and thoughts, are measured by the state of body and mind. Partiality by employers may be observed by violations of the privacy of the state of body and mind.
3. **Privacy of social life**. A person's social life includes the contents of social interactions, such as what was said in a conversation with others on social media platform, and about interactions with a person, when, and for how long. Violating social privacy allows inferences about other types of privacy, e.g., interactions with political groupings can reveal information about a person's opinion.
4. **Privacy of behavior and action**. A person's habits, hobbies, actions, and purchase patterns are considered within privacy of behavior and actions. When shopping online or using credit cards, potentially intimate details are shared with retailers. Exploiting this information for other purposes, such as targeted advertisements, can constitute a violation of privacy.
5. **Privacy of media**. The privacy of a person's images, video, and audio is considered privacy in media, including any CCTV footage uploaded to the Internet. Redistributing or creating user-related media without agreement constitutes a privacy violation.

5.8.3 BUILDING BLOCKS FOR PRIVACY PROTECTION

Building blocks for privacy enhancing techniques (PET) have been developed in recent years. Smart city applications work as a combined range of enabling technologies. PETs are used to protect privacy properties. The type of privacy that PETs protect depends on the context in which it are used, for example, location privacy in smart mobility, and privacy of body and mind in smart health. Privacy properties are independent of the context.

These are six key privacy properties:

1. Anonymity
2. Unlinkability
3. Undetectability
4. Unobservability
5. Pseudonymity
6. Identity management

Anonymity; a subject that is not identifiable within a set of subjects.
Unlink ability: property where two actions or individual's activity cannot be linked by a hacker.

80   Cybersecurity

Undetectability: an attacker cannot sufficiently differentiate whether an item of interest exists.

Unobservability: requires the undetectability and anonymity of all entities involved

Pseudonymity: identifiers instead of original names; identity management refers to the managing of partial identities.

5.8.4 PRIVACY TECHNIQUES

5.8.4.1 Process-Oriented Privacy Protection

Process-oriented privacy protection can be defined and classified as follows:

Privacy by design: Privacy by design includes seven principles that should be followed:

a) Proactive privacy protection, instead of remedial action after privacy violations have happened.
b) Privacy as the default setting.
c) Privacy embedded into the design.
d) Full functionality with full privacy protection.
e) Privacy protection through the entire lifecycle of the data.
f) Visibility and transparency.
g) Respect for user privacy.

Privacy Requirements Engineering: Privacy requirements process shows eight privacy design strategies in two parts.

a) In the first part, four strategies are included regarding data: minimize, hide, separate, and aggregate
b) In second part, four strategies are included regarding process: inform, control, enforce, and demonstrate

Testing and Verification: Privacy ensures testing and verification that the design and implementation of a system do indeed fulfill its privacy requirements.

Transparency: For increased transparency, the provider should openly communicate what type of data is collected, which type of data is stored, how it is processed, who it is shared with, and how this data is protected. Transparency may increase the level of trust for users.

Auditing and Accountability: Accountability means to hold the city accountable for its use of citizens' data and compliance with its privacy policies, e.g., to make sure that citizens pay the correct amounts for usage of public transport, toll roads, or energy. Independent audits also allow the public to understand how, and how often, privacy-invasive technologies are being used, whether they are being used for their stated purpose, and how well they fulfill their purpose.

Privacy Architectures: Privacy architectures are very important to bring the different protections together and ensure that there are no privacy leaks at any point.

Privacy architectures provide access control for centralized storage and merge different cryptographic techniques to provide privacy.

5.8.4.2 Data-Oriented Privacy Protection

Data Minimization: Data minimization reflects the fact that the sensors of modern smart systems naturally gather more sensor data than is required for the task envisioned. This type of data is called collateral data.

5.9 CONCLUSION

The smart city and mobility are relatively new concepts and are always in progress. This chapter focuses on the issues related to the technology implementation and operational issues. Since the topic is related to people, data security is another aspect about which key points are discussed. Government and regulatory authorities need to manage vendors and service providers, and hence need regulatory authority. There is a focus on the status in the various countries of efforts by the government to control smart city projects. Since this is related to technology, many new services may be provided in the coming days.

REFERENCES

Aletà, N. B. (2016). *Smart Mobility in Smart Cities. ResearchGate Publication* (p. 12). València: *CIT2016 – XII Congreso de Ingeniería del Transporte*. Analysis of Policies and Regulations for a. (2020), (p. 9).

Bellavista, P., Corradi, A., Foschini, L., & Ianniello, R. (2015). Scalable and Cost-Effective Assignment of Mobile CrowdsensingTasks Based on Profiling Trends and Prediction The ParticipAct Living Lab Experience. *Sensors*, *15*(8), pp. 18613–18640. ISSN 1424-8220, www.mdpi.com/journal/sensors.

Davor Brčić, M. S., Slavulj, M., Šojat, D., & Jurak, J. (2018). *The Role of Smart Mobility in Smart Cities. Fifth International Conference on Road and Rail Infrastructure CETRA*, (p. 6). Zagreb, Croatia: University of Zagreb.

Deshpande, K. V. & Rajesh, A. (2017). Investigation on IMCP Based Clustering in LTE-M Communication for Smart Metering Applications. *Engineering Science and Technology, An International Journal*, *20*(3), 944–955. *Elsevier*, (p. 12).

Docherty, I., Marsden, G., & Anable, J. (2018). The Governance of Smart Mobility. *Transportation Research Part A: Policy and Practice*, *115*, 114–125. *Elsevier*, (p. 12).

Eckhoff, D., & Wagner, I. (2017). Privacy in the Smart City: Applications, Technologies, Challenges, and Solutions. *IEEE Communications Surveys & Tutorials*, *20*(1), 489–516, (p. 28).

Elmaghraby, A. S. & Losavio, M. M. (2014). Cyber Security Challenges In Smart Cities: Safety, Security and Privacy. *Journal of Advanced Research*, *5*(1), 491–497. (p. 7).

Gebresselassie, M. & Sanchez, T. W. (2018). "Smart" Tools for Socially Sustainable Transport: A Review of Mobility Apps. *Urban Science*, *2*, 45. *MDPI*, (p. 10).

Gumbo, T. & Moyo, T. (2020). Exploring the Interoperability of Public Transport Systems for Sustainable Mobility in Developing Cities: Lessons from Johannesburg Metropolitan City, South Africa. *Sustainability*, *12*(15), 5875. *MDPI*, (p. 16).

Ismagilova, E., Hughes, L., Rana, N. P., & Dwivedi, Y. K. (2020). Security, Privacy and Risks Within Smart Cities: Literature Review. *Information Systems Frontiers*, 1–22. Doi: https://doi.org/10.1007/s10796-020-10044-1

Lim, H. S. M., & Taeihagh, A. (2019). Algorithmic Decision-Making in AVs: Understanding Ethical and Technical Concerns for Smart Cities. *Sustainability*, *11*(20), 5791. *MDPI*, (p. 28).

Luque-Vega, L. F., Carlos-Mancilla, M. A., Payán-Quiñónez, V. G., & Lopez-Neri, E. (2020). Smart Cities Oriented Project Planning And Evaluation Methodology Driven By Citizen Perception—IoT Smart Mobility Case. *Sustainability*, *12*(17), 7088. *MDPI*, (p. 18).

Orlowski, A. & Romanowska, P. (2018). Smart Cities Concept: Smart Mobility Indicator. *Cybernetics And Systems: An International Journal*, *50*(2), 118–131, (p. 23).

Rahmayanti, H., Oktaviani, V., & Syani, Y. (2018). *The Implementation of Smart Trash as Smart*. E3S Web of Conferences, (p. 5). Indonesia: Universitas Diponegoro.

Samih, H. (2019). Smart Cities and Internet of Things. *Journal of Information Technology Case and Application*, *21*, 3–12. (p. 11).

Sarkar, D. A. (2020, November). *Smart City Journals*. Retrieved November 11, 2020, from https://www.thesmartcityjournal.com/en/articles/1333-smart-cities-futuristic-vision

Shabaan, M., Arshad, K., Yaqub, M., Jinchao, F., Zia, M. S., Boja, G. R.,... Munir, R. (2020). Survey: Smartphone-Based Assessment of Cardiovascular Diseases Using ECG and PPG Analysis. *BMC Medical Informatics and Decision Making*, *20*(1), 1–16. (p. 16).

Sumalee, A., & Hung, H. W. (2018). Smarter and More Connected: Future Intelligent Transportation System. *IATSS Research*, *42*(2), 67–71. (p. 5).

Tayyaba, S., Ashraf, M. W., Alquthami, T., Ahmad, Z., & Manzoor, S. (2020). Fuzzy-Based Approach using IoT Devices for Smart Home to Assist Blind People for Navigation. *Sensors (Basel)*, *20*(13), pp. 1–12.

Van Zoonen, L. (2016). Privacy Concerns in Smart Cities. *Government Information Quarterly*, *33*, 472–480. *ELSEVIER*, (p. 9).

Winkowska, J., Szpilko, D., & Pejić, S. (2019). Smart City Concept in the Light of the Literature Review. *sciendo*, *11*(2), 70–86. (p. 17).

Yu, Z., Jin, D., Zhai, C., Ni, W., & Wang, D. (2020). Internet of Vehicle Empowered Mobile Media: Research on Mobile-Generated Content (MoGC) for Intelligent Connected Vehicles. *Sustainability*, *13*(6), 3538. *MDPI*, (p. 21).

6 Digital Forensics Cryptography with Smart Intelligence

Samaya Pillai, Venkatesh Iyengar and Abhijit Chirputkar
Symbiosis International (Deemed University), India

CONTENTS

6.1	Introduction	84
6.2	History of Forensics	84
6.3	Need for Digital Forensics	85
6.4	Sequence of Steps in Digital Forensics	86
6.5	Types of Digital Forensics	86
6.6	The Previous Decade (2001–2011)	87
6.7	Recent Years (2011–2021)	87
6.8	Models of Digital Forensics	88
	6.8.1 The Digital Forensic Research Workshop (DFRWS) 2001	88
	6.8.2 Forensic Process Model (2001)	89
	6.8.3 Abstract Digital Forensic Model (2002)	89
	6.8.4 Integrated Digital Investigation Process Model (IDIP) 2003	89
	6.8.5 Enhanced Digital Investigation Process (2004)	89
	6.8.6 Extended Model of Cybercrime Investigation	89
	6.8.7 Case-Relevance Information Investigation (2005)	90
	6.8.8 Digital Forensic Model Based on Malaysian Investigation Process (2009)	90
	6.8.9 Systematic Digital Forensic Investigation Model SRDFIM (2011)	90
6.9	Real-Time Use Cases of Digital Forensics Application	90
6.10	Summary of Open Source Tools and Techniques Used in Digital Forensics	97
	6.10.1 San SIFT	97
	6.10.2 ProDiscover Forensic	97
	6.10.3 Volatility Framework	98
	6.10.4 The Sleuth Kit (Autopsy)	98
	6.10.5 CAINE	98
	6.10.6 Xplico	98
	6.10.7 X-Ways Forensics	98
6.11	Short Summary of a Few Other Digital Forensic Tools	99

DOI: 10.1201/9781003145042-6

84 Cybersecurity

6.12 Cryptographic Algorithms in Digital Forensics 100
    6.12.1 File Carving Technique 100
    6.12.2 Reconstructing Compressed Data 100
    6.12.3 Recovering Files 101
    6.12.4 Reverse Engineering 101
    6.12.5 Image Integrity 101
6.13 Conclusion 101
References 101

6.1 INTRODUCTION

Industry 4.0 has brought automation to many domains. In this regard, computer hardware devices and software applications have witnessed many new innovations. Most devices have reduced their size but increased their storage capacity and operational speed. The Internet of Things (IoT) and Internet of Everything (IoE) technologies have provided a boost to the implementation of Industry 4.0 in industry domains, including vehicle automation, health care devices, the agricultural sector, and global supply chain operations. Business organizations are becoming more global and paperless in performing routine operations. Similarly, tools have been developed for retrieving and reading digital information from these hardware devices. Conversely, criminal and the terroristic activity rates have also seen an upsurge since the 9/11 episode. In order to counter these challenges, as a counterstrategy newer techniques and intelligence have become acutely essential. Forensics as a field has been on the horizon for quite a long time, and now, digital forensics has become inevitable and a thriving field.

6.2 HISTORY OF FORENSICS

The field of forensics has exited for more than a century. More recently, "digital forensics," earlier called "computer forensics," has been mainly tinkered with by people from law enforcement departments, who have a knack of working with computers/computing technologies. The first noteworthy official mention of digital forensics can be traced to a task force created by the Federal Bureau of Investigation (FBI) in the United States of America in 1984. Known as the "CART" (Computer Analysis and Response Team), its preliminary task was to assist in computer-related investigations. In the subsequent year, in the United Kingdom, a similar task force was established, code-named "Fraud Squad." Starting in 1993, numerous formal conferences were held in the USA and UK to discuss the requirements, process and procedures, and standardization of forensics, as well as other related techniques. These conferences were a collaboration between scholars, experts, law enforcement officers, government organizations, and other stakeholders. In one such conference in the UK, the modern British digital forensic methodology was established. Following this, in 1998, the Scientific Working Group on Digital Evidence (SWGDE) was established, to formulate best practices and standards in digital forensics. Also in 1998, the Association of Chief Police Officers (ACPO) personnel of the UK developed a guide for digital evidence. Called ACPO guidelines, it was primarily followed

Digital Forensics Cryptography

for digital forensics purposes by the law enforcement team. Other organizations, such as the American Society of Crime Laboratory Directors (ASCLDs), also collaborated with it. In 2000, the first Regional Computer Forensic Laboratory (RCFL) was established by FBI, in the USA. In 2002, the NPO (National Program Office) was started as a central organization to coordinate with all other organizations.

6.3 NEED FOR DIGITAL FORENSICS

The global digital forensics market is predicted to grow from US$ 4.6 billion in 2017 to about US$ 9.7 billion by 2022, with a Compound Annual Growth Rate (CAGR) of 15.9% during this period.

The need and priority for digital forensics can be associated with advanced and sophisticated technology, its processing capabilities, and improved performance on various digital fronts. In this regard, Moore's law observes and projects that the number of transistors in a dense integrated circuit (IC) doubles about every two years. In accordance with this law, there is shrinkage in the size of the devices and an increase in the size of the storage and speed capacities of the device. Everybody stores data in some form of data storage devices, as they have become inexpensive and highly efficient. If we look at the timeline of storage devices, we find: 1890-Punch Cards 0.08 KB, 1932-Magnetic Drum 48 KB, 1947-Williams-Kilburn Tube 0.128 KB, 1951-Magnetic Tape Drive 231 KB, 1951-Magnetic Core 2 KB, 1956-Hard Disk Drive (HDD) 3750 KB, 1967-Floppy Disk 80 KB, 1982-Compact Disc (CD) 700,000 KB, 1994-Zip Drive 100,000 KB, 1995-Digital Video Disc (DVD) 1,460,000 KB, 1999-SD Card 64,000 KB, 1999-USB Flash Drive 8000 KB, 2003-Blu-ray Optical Disc 25,000,000 KB, 2006-Cloud Data Storage Unlimited KB. Experts estimate that more than 2,700,000,000,000,000,000 KB (2.7 zettabytes) of data exist in the digital universe today.

In a 2020 Webroot Threat Report, it was revealed that cybercrime statistics for 2019–2020 reported 144.91 million new malware samples. In 2018, it was observed that almost 93% of malware was "polymorphic," i.e., changing its form consistently in order to avoid detection. It was also found that approximately 50% of business PCs, as well as 50% of consumer PCs, got infected. By 2020, we now have over 38.48 million new samples.

It is now popularly said that "data is the new oil" and that social engineering, through deep mining into big data sets, could provide crucial information essential for cybercriminals, as well as posing a major cyberthreat to organizations. Research studies indicate that cybercrimes involve the extraction of valuable information about customers, financial transactions, strategic plans, board members or top management, R&D, user/customer passwords, mergers and acquisitions, intellectual property (patented/non-patented), suppliers, and other stakeholders. These leakages of crucial information have also posed serious cyberthreats to organizations emanating from phishing activities, malware, cyberattacks to disrupt or steal funds or even intellectual property, fraud in various forms, spam, internal attacks, espionage, and natural disasters.

On the other hand, as per the recent global statistics revealed in public domain by Statista (2021), there is an ever-growing number of Smartphone users worldwide.

Smartphone users have grown from almost 1.06 billion users in 2012 to almost 3.8 (approx.) expected users by the end of 2021.

This underlines the growing need for digital forensics, as it applies not only to storage media but also to network/Internet connections, mobile devices, IoT devices, and in reality, any device that can store, access, or transmit data. For this requirement we also have a variety of tools both commercial and open source available to us, depending upon the task at hand.

6.4 SEQUENCE OF STEPS IN DIGITAL FORENSICS

Just as forensics carry out a sequence of steps in the detection of a crime or homicide, similarly, digital forensics has mirrored the steps to be carried out in such a scenario. At the first workshop held at DFRWS in the USA, it was tentatively decided that the following basic sequence of steps needs to be carried out:

Identification > > Preservation > > Collection > > Examination > > Analysis > > Presentation > > Decision.

- a. *Identification:* Identification of the crime. The first complaint required to start processing. This involves the detection of profiles, auditing, etc.
- b. *Preservation:* Preservation of first evidence, the chain of custody etc.
- c. *Collection:* This involves the preservation of evidence, approved methods of handling the hardware or software, etc.
- d. *Examination:* This involves the preservation of evidence, validation techniques, the identification and retrieval of hidden data, patterns in data, etc.
- e. *Analysis:* this involves traceability, data mining, etc.
- f. *Presentation:* This involves documentation, interpretation of the evidence and events, etc.
- g. *Decision:* This refers to decision-making.

6.5 TYPES OF DIGITAL FORENSICS

Disk Forensics: Deals with extracting data from storage media by searching active, modified, or deleted files.

Network Forensics: A sub-branch of digital forensics. It is related to monitoring and analysis of computer network traffic to collect important information and legal evidence.

Wireless Forensics: A division of network forensics. The main aim of wireless forensics is to offers the tools need to collect and analyze the data from wireless network traffic.

Database Forensics: A branch of digital forensics relating to the study and examination of databases and their related metadata.

Malware Forensics: This branch deals with the identification of malicious code, to study their payload, viruses, worms, etc.

Email Forensics: Deals with the recovery and analysis of emails, including deleted emails, calendars, and contacts.

Memory Forensics: Deals with collecting data from system memory (system registers, cache, RAM) in raw form and then carving the data from raw dump.

Mobile Phone Forensics: Mainly deals with the examination and analysis of mobile devices. It helps to retrieve phone and SIM contacts, call logs, incoming and outgoing SMS/MMS, audio, videos, etc.

6.6 THE PREVIOUS DECADE (2001–2011)

Mocas (2004) proposes a framework for development, evaluation, and research in digital forensics, primarily focusing upon context, integrity of data, reproducibility, authentication, and noninterference. Other challenges for next-gen digital forensics in this arena include inefficient computing systems, unnecessary delays, and deployment failures (Richard & Roussev, 2006). Research into digital forensics has since attempted to consider data collection audit trails and internal logs for preservation (Bradford et al., 2004); investigation models (Pollitt, 2007); proactive digital forensics (Ray, 2007); software and techniques for virtualization (Pollitt et al., 2008); evidence modeling, data volume, network forensics, media types, live acquisition, and control systems (Nance et al., 2009); the detailed quantitative analysis of specified digital media formats (Turnbull et al., 2009). Until the end of 2010, digital forensics thrived in a very rudimentary form, in the "golden age of digital forensics" as acclaimed by Simson L. Garfinkel (2010); yet, most of those tools, research studies, and processes were growing obsolete by then. There was dire need for dramatic improvements in these tools and research processes to achieve operational efficiencies, while presenting forensic data and implementing forensic computations. Standardization was felt necessary while creating file system metadata, ontologies, and schema to support digital forensics. Forensic tools are vital, because they have the ability to recover data and reconstruct crucial evidence that could reside in computers and such digital media, for solving crimes committed with computers, including those against people. Earlier digital forensic techniques were focused upon network and memory forensics, to investigate evidence found on time sharing systems, computer hacks, residual data, recovery of emails/instant messages, and storage devices, as well as recovering deleted files, and implementing simple file carving.

6.7 RECENT YEARS (2011–2021)

Since 2010, there has been a growing crisis in digital forensics, owing to increased challenges and advancements in technology with respect to the size of storage devices; embedded flash storage and memory devices; the complexity of file formats; multiple operating systems; proprietary systems and apps; open source tools; tools for data exploitation; encryption; remote cloud processing; malware; viruses, and other reasons. Much more than that, the challenge is great to have a standardized process to implement digital forensics. Parallel to this challenge is that encryption and cloud computing architectures pose a threat to forensic visibility, in turn denying access to ingrained data for further investigation and limiting available functionalities. It is clear and evident that the available tools and techniques in digital forensics face severe challenges while dealing with data evidence, complex digital and computer systems, myriad data volumes, residual memories, cross-drive analysis, and cross-case searches, and thus are poorly suited to handle manual forensic investigations or interventions.

6.8 MODELS OF DIGITAL FORENSICS

As the domain of digital forensics has grown, models have been conceptualized by various thinkers and scientists, as per their requirements and perceptions. These models include:

6.8.1 THE DIGITAL FORENSIC RESEARCH WORKSHOP (DFRWS) 2001

This was one of the first workshops conducted on the topic. The people involved were from all disciplines, from academics, to law enforcement, and the military. The major outcome of this workshop was the development of the outline of the digital forensics process. The personnel attending the workshop were divided into four groups, and each group discussed a pertinent point related to digital forensics. At the end, they presented it to all the attendees. The first group, in Workshop 1, was assigned to "Define a Framework for Digital Forensic Science." The second group, in Workshop 2, was assigned to "Discuss the Trustworthiness of Digital Evidence." The third group, in Workshop 3, was assigned to "Discuss Detection and Recovery of Hidden Data." The fourth group, in Workshop 4, was assigned to "Discuss Digital Forensic Science in Networked Environments Network Forensics."

The first group came up with a formal definition of digital forensic science.

> The use of scientifically derived and proven methods toward the preservation, collection, validation, identification, analysis, interpretation, documentation and presentation of digital evidence derived from digital sources for the purpose of facilitating or furthering the reconstruction of events found to be criminal, or helping to anticipate unauthorized actions shown to be disruptive to planned operations (Palmer, 2001).

They also defined the process to be followed in carrying out digital forensics, which was similar to the process carried out in a normal criminal investigation.

The second group came up with suggestions regarding the trustworthiness of digital evidence in the court of law. Since the nature of digital evidence is abstract, it has many layers. The integrity of the data can be compromised before it is formally produced as evidence. Because it needs to be analyzed initially, while doing so, it can easily be modified or transformed. Also, there were no standards as such during this time. Hence, as a precaution, it was suggested that the analysis needs to be done on a copy or clone of the digital device.

The third group focused on the detection and recovery of hidden data. Techniques employed to hide data were abundant, and identifying them was a challenge.

Digital Forensics Cryptography

Steganalysis and Steganography were discussed. This is defined as "the practice of concealing a file, message, image, or video within another file, message, image, or video." Watermarking, hashing, encryption, and decryption techniques were also discussed.

The fourth group discussed digital analysis in networked environments. This was a vast and growing field. It was decided that the entire framework required for the investigation needed to be developed keeping in view the time, complexity, and tools of the networking evidence.

6.8.2 FORENSIC PROCESS MODEL (2001)

The U.S National Institute of Justice (NIJ) developed a document for first responders. The document served as a guide for the law enforcement and other responders who have the responsibility for protecting an electronic crime scene and for the recognition, collection and preservation of digital evidence (Ashcroft, 2001).

6.8.3 ABSTRACT DIGITAL FORENSIC MODEL (2002)

Researchers Reith, Carr and Gunsch (2002) analyze forensic models and propose a framework using traditional physics forensics as the baseline.

They suggest the following steps:

Identification > > preparation > > strategy of approach > > preservation > > collection> > examination > > analysis > > presentation > > returning of evidence.

6.8.4 INTEGRATED DIGITAL INVESTIGATION PROCESS MODEL (IDIP) 2003

Carrier and Spafford (2003) propose this model, with a review of their earlier work, and map the digital investigation process onto the actual physical process of investigation.

6.8.5 ENHANCED DIGITAL INVESTIGATION PROCESS (2004)

Baryamureeba and Tushabe (2004) suggest a modification to Carrier and Spafford's Integrated Digital Investigation Process Model (2003). Their proposal adds two more phases to the original model. They suggest that crime scenes can be classified into the primary crime scene, i.e., the computer, and the secondary crime scene, i.e., the physical crime area. According to their theory, this distinction could avoid inconsistencies in an investigation.

6.8.6 EXTENDED MODEL OF CYBERCRIME INVESTIGATION

This model is proposed by Ciardhuáin (2004), which focuses upon the "management" attribute. The model gives a basis for comprehending the overall process of investigation and captures most of the information.

6.8.7 Case-Relevance Information Investigation (2005)

Ruibin, Yun, and Gaertner (2005) propose this model introducing two terms: "Seek Knowledge" and "Case Relevance." They also suggest that a computer's intelligence could be used in the investigation process. This is something close to using Artificial Intelligence (AI) to solve a case or across cases. They also differentiate between computer security and forensics.

6.8.8 Digital Forensic Model Based on Malaysian Investigation Process (2009)

Perumal (2009) describes this model. He suggests that once an investigation has started, the evidence can be classified into live data and static data.

6.8.9 Systematic Digital Forensic Investigation Model SRDFIM (2011)

Agarwal et al. (2011) propose this model consisting of 11 phases. Their objective is to set up policies and procedures that could help in a systematic investigation process. Their main area of interest is limited to cybercrimes and computer fraud.

6.9 REAL-TIME USE CASES OF DIGITAL FORENSICS APPLICATION

USE CASE 1: IDENTIFICATION OF LOCATIONAL DATA EMBEDDED IN GEOTAGS USEFUL IN DIGITAL FORENSIC INVESTIGATIONS

Primary actor - Army soldier.

Secondary actor(s) - Smartphone device, geotag information, social media account.

Scope and goals - Allegations about Russian Army operating in foreign territory following Russian annexation of Crimea, Ukraine, in February 2014.

Level - Strategic.

Stakeholders and interests -

a. Sergeant Alexander Sotkin from the Russian Army posted multiple selfies for over a month-long time-period from his cell phone into his own public media accounts, such as Instagram.

b. The press discovers the posted media files (.jpeg), which were also embedded with geotag metadata.

c. The Ukraine military alleged Russian Army involvement in Crimea without authority.

Precondition - International border tensions and border crossing; denial from Russian Army about activity in Ukrainian territory.

(Continued)

Digital Forensics Cryptography

Minimum guarantee - Geotagged files contain locational metadata useful for digital forensic investigations.

Success guarantee - Geotagged files with location metadata reveal/confirm Russian Army soldier's movements from Russian military base into eastern region of Ukraine and return to base.

Trigger - Geotagged image confirms/reveals locational metadata.

Key success scenario -

1. Geotags generated by smartphones provide much precise and accurate locational metadata.
2. Locational metadata can be used for further legal and digital forensic investigations.

Extensions -

1a. Geotagged locational metadata includes geographic positions, timelines, and timestamps mapped into the smartphone apps residing with the image object.
1b. Geotagged locational metadata can also be embedded with other file formats, such as video files, SMS text messages, etc.
2a. Legal investigation could be conducted to extract relevant information about who, what, when, where from geographic timelines and timestamps datasets.
2b. Locational metadata could be further extracted from linked applications, WiFi network connections, mobile network tower logfiles, network provider databases (call history, logs, and recordings, etc.)

USE CASE 2: DIGITAL FORENSIC INVESTIGATION OF INCRIMINATING EVIDENCE TO PROVE INVOLVEMENT OF KRENAR LUSHA, UK, IN CRIMINAL/TERRORISM ACTIVITIES

Primary actor - Krenar Lusha, United Kingdom.

Scope and goals - Search for digital evidence tantamount to terrorism activities based upon technological usage involving Internet search patterns found in laptop and its incidental material evidence.

Level - Summary

Stakeholders and interests -

a. Krenar Lusha from the United Kingdom, suspected of criminal/terrorist activities, was duly arrested.

(Continued)

b. Police conducted criminal investigation evidences from laptop and Internet usage logs and downloaded materials, explosive materials at apartment, and chat messages via MSN

Precondition - Arrested for suspected criminal/terrorist activities and existence of incriminating material

Minimum guarantee -

1. Laptop contains digital evidence tantamount to criminal usage
2. Police recover incriminating documents and associated materials (physical)

Success guarantee - Internet search patterns, downloaded online manual for 4300 GM explosives and search belts, recovery of arms and ammunition stores for making explosives collaborate with criminal activities

Trigger - Investigation of Internet search logs and patterns correlates incriminating documents, materials, and associated activities

Key success scenario -

1. Usage of laptop with Internet facility to search for useful information associated with criminal activity, chat messages, and self-presentation over online media
2. Further legal investigation through search of residential location helps recover relevant quantities of petrol, potassium nitrate, and live shotgun cartridge materials
3. Building legal evidences to prove criminal intentions of subject

Extensions -

1a. Online search and download of manual for 4300 GM explosives and search belts
1b. Internet search patterns and logs, chat messages over MSN, and electronically downloaded documents correlate with criminal intentions
2a. Explosives manufactured from such associated materials could be used for criminal activities
3a. Criminal intentions corroborated from Krenar Lusha's chat messages over MSN, all presenting himself as a terrorist/sniper with the objective purpose to kill Jewish and American people. This provides strong incriminating evidence that could be used in a court of law to prove guilt

Digital Forensics Cryptography

USE CASE 3: DATA FROM ASSET TRACKERS – SENSORS AND IOT DEVICES (CASE OF HOWZE VERSUS WESTERN EXPRESS INC.)

Primary actor - a) Howze's asset trackers with advanced technology. b) The injured motorcycle victim hit by tractor-trailer. c) The hit-and-run accused suspected to be from Western Express Inc.

Scope and goals - Structured data analysis in digital forensics performed using asset tracker systems to extract relevant evidence to prove an act of vehicular accident leading to a criminal act

Level - Summary

Stakeholders and interests -

   a. Motorcycle victim

   b. Tractor-trailer driver

   c. Howze's asset tracker system

Precondition –

   a) Victim was riding a motorcycle and was knocked off the road by a tractor-trailer

   b) The suspected hit-and-run accused caused injury to the motorcycle victim

   c) The eyewitness only recalls that the tractor-trailer logo read "Western Express".

Minimum guarantee -

   1. The motorcycle victim suffered accidental injuries

   2. The accused forced a motorcycle rider off the road while riding a tractor-trailer

Success guarantee - Only if the eyewitness identification statement that the tractor-trailer that read "Western Express" logo actually was involved in the accident turns out to be proven

Trigger - Asset tracker with GPS sensor features, including WiFi and Bluetooth, reveals locational data including expanded features to establish the presence of such moveable assets at the accident spot successfully

Key success scenario -

   1. Asset tracker system comprehensively and accurately provides locational data (geographical) using GPS, WiFi, and Bluetooth capabilities

   2. Defendant's trucks equipped with such asset trackers were serviceable and actually recorded the locational dataset into the centralized database

(Continued)

94   Cybersecurity

Extensions -

1a. Locational data are stored and retrieved from centralized and secure databases, wherein Western Express' six-month GPS data is retained, as per policy requirements. Also, expanded features could provide relevant information incidental to proving the accident

2b. Material evidence provided from the asset tracker system was challenged for correlation between six-month data retention policy and investigation conducted 27 months after the accident

USE CASE 4: XIAOLANG ZHANG - CORPORATE TRADE SECRETS THEFT AND NETWORK FORENSICS

Primary actor - Xiaolang Zhang, Engineer for Apple's autonomous car division

Scope and goals - Investigation into Zhang's network activity for data search over specific timelines reveals intense bulk searches, targeted downloading of information from company's secret databases

Level - Summary

Stakeholders and interests -

a. Xiaolang Zhang and his engagement with corporate network, involving unauthorized access to corporate information from secret databases

b. Company manager intrigued with Zhang's resignation decision and unauthorized network activities

Precondition -

a) Xiaolang Zhang accessed company information from secret databases without valid access privileges or permission

b) Xiaolang Zhang suddenly announcees resignation from work and returns to China

c) Zhang is indicted for theft of trade secrets

Minimum guarantee -

1. Theft of trade secrets from secret databases of the company

2. Unauthorized access into company's network systems

3. Bulk searches and targeted downloading copious pages of information establishes Zhang's network activity

(Continued)

Digital Forensics Cryptography

Trigger - Zhang's network activity is highly alarming, leading to unauthorized access to trade secrets

Key success scenario -

1. Network forensics help retrieve log data from network servers, and networking tools such as firewalls, routers, intrusion detection applications
2. Network forensics assist in network activity tracing and monitoring history to establish intrusion and breach detection, in order to prove the network breach committed by Zhang

Extensions -

1a. Availability of network data presents opportunities for investigation into user activities in non-cyber cases (such as theft of trade information/ secrets)
2a. Network logs are analyzed to identify suspicious user activity, including massive engagements with network data, tampering, and associated transactions with the database

USE CASE 5: DENNIS RADER - THE SERIAL KILLER

Primary actor - Dennis Rader, serial killer in Kansas in 2005

Scope and goals - Establishing linkages of Dennis's MS-Word file (deleted) containing the letter with the established crime

Level - Summary

Stakeholders and interests -

a. Dennis Rader, serial killer in Kansas in 2005
b. The victim
c. The police

Precondition -

a) Floppy disk sent to the police contained a letter
b) Disk forensic investigators find a deleted MS-Word file with metadata revealing "Dennis" as the last person to modify the deleted file
c) The metadata indicates some link between the deleted file and the Lutheran church where Radar was a deacon

(Continued)

Minimum guarantee -

1. The deleted file in the floppy disk contains metadata in a MS-Word deleted file, which reveals associated information and link with "Dennis"
2. Dennis Radar actually sent the police a floppy disk with a letter on it

Trigger - A deleted MS-Word file provides metadata containing the name "Dennis," as the last person to modify the deleted file, and a link to a Lutheran church

Key success scenario -

1. Metadata from deleted MS-Word file provides user activity
2. The metadata names "Dennis" as the last person to modify the deleted file

Extensions -

1a. File's metadata actually reveals the last user name as "Dennis"

2a. Since the name of "Dennis" is found in the floppy disk file metadata, his involvement could be linked

USE CASE 6: SCOTT TYREE AND ALICIA KOZAKIEWICS - ESTABLISHING A CASE OF KIDNAPPING

Primary actor - Scott Tyree, the kidnapper

Scope and goals - Tracing the kidnapper using Yahoo! physical address of subscriber

Level - Summary

Stakeholders and interests -

a. Scott Tyree, the kidnapper
b. Alicia Kozakiewicz, the 13-year-old kidnapped
c. Another person who provided FBI with instant message content
d. Yahoo! IP Address

Precondition -

a) Scott Tyree kidnapped and imprisoned 13-year-old Alicia Kozakiewicz
b) Scott sent an instant message of the photograph with the Yahoo! screen name "masterforteenslavegirls," showing Kozakiewicz to another man
c) FBI contacted Verizon to establish the name and physical address of the subscriber to whom the IP address was assigned

(Continued)

Digital Forensics Cryptography

Minimum guarantee -

1. The instant message with a photograph showing Kozakiewicz is sent by Tyree
2. The other person provides the FBI with the Yahoo! screen name of the person who sent the message. "masterforteenslavegirls" was established by Verizon to link with the name of the person to whom the IP address was assigned, as Scott Tyree

Trigger - The Yahoo! screen name, the message "masterforteenslavegirls," and the IP address of assigned subscriber link to Tyree

Key success scenario -

1. Verizon is able to establish the name and physical IP address of subscriber
2. Scott Tyree sent an instant message with photograph showing Alicia Kozakiewicz

Extensions -

1a. The assigned subscriber to the physical IP address is Scott Tyree
2a. The instant message links with the Yahoo! screen name, and Verizon verified the physical IP address subscriber to be that of Scott Tyree
2b. The linking details establish the connection of Scott Tyree with the crime of kidnapping Alicia Kozakiewicz

6.10 SUMMARY OF OPEN SOURCE TOOLS AND TECHNIQUES USED IN DIGITAL FORENSICS

Computer forensics are mainly concerned with the tools used in the process of forensics. The tools have been evolving for years. Some of the best tools are discussed and few more are listed.

6.10.1 SAN SIFT

The SANS investigate forensic toolkit (SIFT): this is a Linux based tool which helps in conducting in-depth forensic with open-source tools that are updated frequently. It has the timeline from system logs, a Scalpel for data file carving, and Rifiuti for investigating the recycle bin. By the use of these tools, it can support various witness formats.

6.10.2 PRODISCOVER FORENSIC

This is a powerful computer security tool. It enables us to locate all data on a computer disk. It can also be used to protect evidence and develop evidentiary reports to be used in legal proceedings. It can be used to recover deleted files, investigate slack space, investigate alternate data streams, investigate hardware-protected areas of the disk, and since it reads at sector levels, it can identify any hidden data. Key features of ProDiscover Forensic include that it can create a complete bit stream copy of the disk and it can access the entire disk for data, including hidden and deleted files, along with metadata.

6.10.3 VOLATILITY FRAMEWORK

This mainly focuses on the analysis of memory based on the analysis of data found in RAM. Since RAM is volatile, it could affect runtime or dynamic system analyses. It can also do a cross-platform, extensible investigation. It is widely used in academic and commercial domains for digital forensics.

6.10.4 THE SLEUTH KIT (AUTOPSY)

This is a command line tool and is open source. It is mainly used to analyze a disk and recover data. It analyzes in particular the volume and file systems. It has a plug-in feature that can upscale the modules used in investigation. It has a GUI interface and can also be used in investigating smartphones. The plug-in mode can be used to develop add-on modules depending on the requirements, using the programming languages Java or Python. It can be used by multiple-users concurrently. It can graphically trace events in the system. It supports NLP-related activities, such as keyword search, text extraction, and web–artifact extracts. It can trace and identify shortcuts, identify recently accessed documents in the system, and extract geological details from the image file in the system.

6.10.5 CAINE

This stands for Computer Aided Investigative Environment. Based on the Linux operating system, it has a variety of forensic tools and a user-friendly interface. This interface enables interaction with many open source forensic tools. The current version meets the standards of forensic safety and reliability. It can perform network forensics, as well as mobile forensics. It can also help in data recovery.

6.10.6 XPLICO

This network forensics tool can reconfigure and reconstruct data from packages such as wireshark, tcpdump, etc. It can do this with web pages, image files, data files, etc. It can support and communicate across protocols of the application layer and network layers, including HTTP, SIP, IMAP, POP, SMTP, TCP, UDP, IPv4, and IPv6. It supports multi-threading and also databases such as MySql and SqlLite. It does not have size limitations as such.

6.10.7 X-WAYS FORENSICS

This is based on the Windows operating system. Its significance is its very light, it need not be installed, and it can run off a USB. Used mainly in tracing deleted files

Digital Forensics Cryptography

and data, it can be used in the cloning of a disk. It can read raw file structures of images, as well as large-scale data. It can edit binary data.

6.11 SHORT SUMMARY OF A FEW OTHER DIGITAL FORENSIC TOOLS

WindowSCOPE: A tool focused on memory forensics. It helps in analyzing a volatile memory. It supports in examining the windows kernel, DLLS, the virtual as well as physical memory.

Encrypted Disk Detector: Used to detect encrypted drives. It supports TrueCrypt, PGP, Bitlocker, and Safeboot encrypted volumes.

Wireshark: A tool used to monitor network data. It can capture, analyze, and report events and incidents in the network.

Magnet RAM Capture: A tool used to copy and monitor the physical memory of a computer and analyze it. It supports the Windows operating system.

Network Miner: A tool used to investigate a network. It has a user interface. It can be used across different operating systems, such as Windows, Linux, and Mac. It performs packet sniffing and can detect the host name, OS, session, and open ports.

NMAP: NMAP (Network Mapper): Used in network forensics. It supports all operating systems—Windows, Linux, and Mac.

RAM Capturer: Used to store the memory dump of the volatile memory of computer. Memory dumps may contain passwords, log-in details of websites, mail, social network sites, etc.

Forensic Investigator: A Splunk tool that has a combination of the formerly listed tools.

FAW (Forensics Acquisition of Websites): Used to capture web pages. The content from these include web pages, images, and source code.

HashMyFiles: Calculates hash values using the algorithms MD5 and SHA1.

USB Write Blocker: Used to analyze USB drives without leaving behind traces such as fingerprints, timestamps, or changes to the metadata.

Crowd Response: A Windows application to gather system information for incident response and security engagements. It can export the results to any file with the extensions XML, CSV, HTML, etc., which are easy for an end-user to operate. Additionally, an open source tool such as Tortilla enables the secure, anonymous, and transparent routing of all TCP/IP and DNS traffic.

Shellshock Scanner: Scans a network for any shellshock vulnerability.

Heartbleed scanner: Scans a network for OpenSSL heartbleed vulnerability.

NFI Defraser: Detects full and partial multimedia files in data streams.

ExifTool: A tool to read, write, and edit meta information. It works for different file types. It also works with graphics files. It can read EXIF, GPS, IPTC, XMP, JFIF, GeoTIFF, Photoshop IRB, FlashPix, etc.

100 Cybersecurity

Toolsley: Has more than 10 useful tools for investigation, such as a password generator, text encoder, hash generator, file identification, file verification, etc.

Volatility: A forensic tool used for memory-related investigations. It can capture information about processes, sockets, DLL, and registry. It can also capture information from dump files, hibernation files, crash files, etc.

Dumpzilla: Extracts and analyzes information from browsers such as Firefox, Iceweasel, and Seamonkey.

Browser History: Foxton has two tools available free of cost. (a) *Browser history capturer*, which can extract and capture the information and history from web browsers including Firefox, Chrome, and Internet Explorer. It works well with Windows OS. (b) *Browser history viewer* that can capture and analyze the web history from all modern browsers. It can also reflect the information in graphs. There is a facility to filter the data.

ForensicUserInfo: Captures details such as hash, password details, password resets, account expiration, log-in details, profile paths, etc.

Kali Linux: A distro of the Linux operating system. It is used for penetration testing and has forensic features.

Paladin: A forensic suite that is famous and used by many. It is a distro of the Linux operating system.

Sleuth Kit: Uses a command line interface. It can be used for file system analysis.

6.12 CRYPTOGRAPHIC ALGORITHMS IN DIGITAL FORENSICS

6.12.1 FILE CARVING TECHNIQUE

This has been one of the most import developments in forensic science. It can recover completely deleted files in the system, i.e., even from the trash or recycle bin. This retrieval is of the deleted files and from their deleted metadata. It was invented around 1999 by independent security researcher Dan Farmer. The technique is widely used globally. In this technique, the file header and file footer are traced. These two sequences and the data within them are traced, identified, and retrieved. The technique can retrieve data files and image files. It can also recreate files that were broken into many pieces, which is a very complex process.

6.12.2 RECONSTRUCTING COMPRESSED DATA

This process reconstructs compressed data. Compression is a widely used process to save space. Text data is compressed with lossless algorithms that support restoring the original data. But images and videos have a lossy system. Currently, these are overcome, and they can also be retrieved and reconstructed. In 2009, a fragment of jpeg was reconstructed even when the file was missing or had been deleted from the system. In 2011, data was reconstructed from files compressed with zip or deflate software.

6.12.3 Recovering Files

The temporary or volatile memory, i.e., RAM, has recent programs and applications running in the system. But the state of RAM changes rapidly. To recover and reconstruct this data is a huge challenge. Recently, this was achieved. Techniques have been developed to acquire and analyze the contents of a running computer system, a process called memory parsing. It is very useful in tracing the damage done by a computer virus or worms and returning the system to an earlier stage. Memory parsing, combined with file carving, can help in recovering digital photographs and video.

6.12.4 Reverse Engineering

Today's techniques to extract allocated files from disk images were largely developed through this method. It helps in back-tracing to an earlier state. We also call this a part of system analysis.

6.12.5 Image Integrity

Now, even images and photos can be recovered. Therein lies a crucial question of image integrity, i.e., validating that the recovered image was the actual image and not a modified version. In the past, there has evidence that photos were tampered with. This is easy with much software, such as Photoshop, CorelDraw.

6.13 CONCLUSION

Today's world provides powerful digital forensic technology. Digital forensics as a functionality not only applies to storage media but also to network, Internet connections, mobile devices, IoT and IoE devices, and in reality, any device that can store, access, or transmit data. As a result, we have a variety of tools, both commercial and open source, available to us, depending on the task at hand. This is a big help in the investigating crimes:, cybercrimes, espionage, blackmail, identity theft, data theft, illegal online activities and transactions, and a plethora of other malicious activities. Cybercrime being such a big business, the response of law enforcement officials and agencies should be equally strong and impressive in their research, development, intelligence, and training divisions, if they are to put up a fight in what may seem like a never-ending battle in the digital world. It still requires lot of manpower and government resources. The existing digital investigation procedures and practices require extensive interaction with humans. The current branch of machine learning and AI can overcome this limitation. It can also help attain faster, more accurate results. With the advancement of technology, the tools for digital forensics must be regularly updated; standard techniques, protocols, and procedures are required in the digital forensics field, in terms of the changing technological scene.

REFERENCES

Agarwal, A., Gupta, M., Gupta, S., & Gupta, S. C. (2011). Systematic digital forensic investigation model. *International Journal of Computer Science and Security (IJCSS)*, 5(1), 118–131.

Ashcroft, J. (2001). Electronic crime scene investigation: A guide for first responders. In *National Institute of Justice (US), Technical Working Group for Electronic Crime Scene Investigation*. Washington, DC: US Department of Justice, Office of Justice Programs, National Institute of Justice. Available at: https://www.ncjrs.gov/pdffiles1/nij/187736. pdf [Accessed January 20, 2021]

Baryamureeba, V., & Tushabe, F. (2004). *The enhanced digital investigation process model. Digital Investigation*. Available at: https://dfrws.org/wp-content/uploads/2019/06/2004_ USA_pres-the_enhanced_digital_investigation_process_model.pdf [Accessed January 20, 2021]

Bradford, P. G., Brown, M., Perdue, J., & Self, B. (2004, April). *Towards proactive computer-system forensics*. In *International Conference on Information Technology: Coding and Computing, 2004 Proceedings. ITCC 2004*. (Vol. 2, pp. 648–652). Las Vegas, Nevada: IEEE.

Carrier, B., & Spafford, E. H. (2003). Getting physical with the digital investigation process. *International Journal of Digital Evidence*, 2(2), 1–20.

Ciardhuáin, S. Ó. (2004). An extended model of cybercrime investigations. *International Journal of Digital Evidence*, 3(1), 1–22.

Garfinkel, S. L. (2010). Digital forensics research: The next 10 years. *Digital Investigation*, 7, S64–S73. DOI: http://dx.doi.org/10.1016/j.diin.2010.05.009

Mocas, S. (2004). Building theoretical underpinnings for digital forensics research. *Digital Investigation*, 1(1), 61–68.

Nance, K., Hay, B., & Bishop, M. (2009, January). *Digital forensics: Defining a research agenda*. In *2009 42nd Hawaii International Conference on System Sciences* (pp. 1–6). Big Island, Hawaii: IEEE.

Palmer, G. (2001, August). *A road map for digital forensic research*. In *Proceedings of the Digital Forensic Research Workshop* (pp. 27–30). Utica, New York: AFRL/IFGB. Available at https://dfrws.org/wp-content/uploads/2019/06/2001_USA_a_road_map_ for_digital_forensic_research.pdf

Perumal, S. (2009). Digital forensic model based on Malaysian investigation process. *International Journal of Computer Science and Network Security*, 9(8), 38–44.

Pollitt, M. M. (2007, April). *An ad hoc review of digital forensic models*. In *Second International Workshop on Systematic Approaches to Digital Forensic Engineering (SADFE'07)* (pp. 43–54). Bell Harbor, WA: IEEE.

Pollitt, M., Nance, K., Hay, B., Dodge, R. C., Craiger, P., Burke, P., … & Brubaker, B. (2008). Virtualization and digital forensics: A research and education agenda. *Journal of Digital Forensic Practice*, 2(2), 62–73.

Ray, D. A. (2007). Developing a proactive digital forensics system. PhD dissertation, University of Alabama, 2007.

Reith, M., Carr, C., & Gunsch, G. (2002). An examination of digital forensic models. *International Journal of Digital Evidence*, 1(3), 1–12.

Richard III, G. G., & Roussev, V. (2006). Next-generation digital forensics. *Communications of the ACM*, 49(2), 76–80.

Ruibin, G., Yun, T., & Gaertner, M. (2005). Case-relevance information investigation: Binding computer intelligence to the current computer forensic framework. *International Journal of Digital Evidence*, 4(1), 1–13.

Statista.com (2021). Number of smartphone users worldwide from 2016 to 2021. Available at: https://www.statista.com/statistics/330695/number-of-smartphone-users-worldwide/ [Accessed January 20, 2021]

Turnbull, B., Taylor, R., & Blundell, B. (2009, March). *The anatomy of electronic evidence–Quantitative analysis of police e-crime data*. In *2009 International Conference on Availability, Reliability and Security* (pp. 143–149). Fukuoka, Japan: IEEE.

7 Transmission Modeling on Malware Attack through IoTs

Yerra Shankar Rao
Gandhi Institute of Excellent Technocrats, India

Binayak Dihudi
Konark Institute of Science and Technology, India

Tarini Charan Panda
Ravenshaw University, India

CONTENTS

7.1 Introduction 103
    7.1.1 Basic Terminology 106
7.2 Hypothesis and Mathematical Model Formulation.....................................106
    7.2.1 Mathematical Model Formulation 107
    7.2.2 Mathematical Model Analysis 108
7.3 Equilibrium Points and Basic Reproduction Number.................................108
    7.3.1 Existence of Stability of the Equilibrium Points.............................109
    7.3.2 Local Stability of the Malware-Free Equilibrium Point 109
    7.3.3 Local Stability of the Endemic Equilibrium 110
7.4 Global Stability of the Equilibrium Points..112
    7.4.1 Global Stability for Endemic Equilibrium Point 113
7.5 Numerical Simulation and Results...114
7.6 Conclusion..117
References..117

7.1 INTRODUCTION

Nowadays, due to the advancement of Internet technologies, the human lifestyle has changed in exceptional ways. People are closely associated with the Internet and its technologies of hardware devices such as smartphones, computers, laptops, and different electronic devices. In recent years, the popularity of the IoT has rapidly increased. So has the number of applications introduced in the domain of the IoT, such as environmental tracking, transportation and manufacturing management, all

DOI: 10.1201/9781003145042-7

branches of defense (Air, Navy, Army), medical science, home automation, lighting fixtures, thermostats, digital cameras, technological farming, etc. Essentially, the IoT means the devices are communicating with wired or wireless technologies through many sensors, such as actuators, circuits, routers, and gateways. The devices collect and send data by using enabled devices. The IoT is the Internet connectivity of physical devices. It is loaded and saved inside the electronic devices through microprocessor chips. After connection, those devices can interact with each other thorough the Internet and can be controlled by the operator through remote sensing devices. Sometimes, the IoT is also called the Internet of Everything. By using embedded sensors, communication hardware, and the processor, the IoT attains data from its surrounding environment. This also supports one or more common ecosystems, such as smartphones, smart speakers, etc., associated and controlled by this device. The IoT plays a major role in medical science, for health monitoring, such as measuring the blood pressure range, to alert emergency notifications systems, and other activities. Aurdino is a device that plays a major role in an IoT network. In this case, software is loaded to write a program and give the command to burn it in this device or microprocessor. This is not only insecure, as it needs an Internet connection, but it can also be easily hacked without the Internet. The use of the Internet is increasing rapidly, and so are the threats against IoT device infrastructure applications. Different types of attack occur in IoT devices: physical attacks, encryption attacks, denial of services attacks, firmware, hijacking, botnets, man in middle concept, data identity theft, social engineering, ransomware, and eavesdropping. Among the varieties of attack is the botnet attack. This happens through malware called the Mirai botnet. This type of attack began in 2016. The malware controls IoT devices for the purpose of creating a botnet and a conducting a DoS attack. The work of Mirai is to scan all the portions of IoT devices and to attempting to login, putting the series of pass words or usernames together, by which the devices are infected and the server targeted with malicious objects used in the software. When Mirai continues to operate in IoT devices, it steals personal information, abuses online banking data, sends phishing emails, and corrupts and slows down popular services, making the device unstable, as well as unrecoverable. Given the rapid growth of Internet technology and its cyberthreats, such as malware propagation and behavior, it is important to protect IoT devices from malware attacks through viruses the size of a botnet.

Therefore, it is necessary to form an IoT-based epidemiological model. Researchers have studied and established protocols for the detection of IoT malware transmission to reduce the threat tosecurity posed by malware code. Gardner et al. [1] propose an IoT BAI (IoT Botnet Awareness Information) model based on the concept of the epidemic model (SEIRS). The model considers the possibility of mitigating the frequency of IoT botnet attacks with improved user information that may positively affect user behavior. The positive behavior and improved user information was developed for mitigation of IoT botnet attacks in the network. Ji et al. [2] give an overview of the architecture of the Mirai botnet and analyze its life cycle. In the analysis phase of malware propagation, an SIR format was used, where N represents the total number of IoT devices in a region. Conclusively, the infected node cannot transmit the infections to other nodes in the network. Acarali et al. [3] analyze IoT botnet formation based on specific characteristics that include energy, limited processing power, and the density of a node to the proposed IoT-SIS model. The findings reveal that maximizing the effectiveness of transmission probability can determine the botnet propagation strategies in IoT networks, rather than concentrating on the contact rate. A few works consider botnet formation in an IoT-based WSN with worm similarities in terms of propagation, whereas the memory and energy efficiency are the main important estimating factors of WSNs, due to the loss of data and network lifetime [4]. C. Kolias et al. [5] provide an overview of Mirai attacks and the commensurate security policies. Y. S. Rao et al. [6] explain the effect of vaccination in the two sources on a distributed denial of service (DDoS) attack in the network. Hezam Akram et al. [7] discuss the guidelines for an IoT-based attack and its defense, theoretically. Y. S. Rao et al. [8] discuss the pre- and post-quarantine defense in the network using an epidemic model. C. Zhang et al. [9] discuss the common defense algorithm for DDoS attacks over IoT devices. M. J. Farooq et al. [10] focus on malware propagation in a wireless network and also give a brief idea of the mitigation of the network dynamically. Mishra et.al [11] explain the automata-based solutions for DDoS attacks on IoTs. Zhang et.al [12] propose challenges and a vision for the IoT. Riahi Sfar et.al [13] contribute their research for making the framework or the roadmap of IoT devices and their security. J. A. Jerkins et al. [14] study the IoT SIS propagation model and consider the impact of some IoT-specific characteristics. Singh et al.[15] discuss applications of the IoT and robotic technology in disaster monitoring, where robots are protecting the IoT devices from attacks and surveillance in the network. Q. Zhu et al. [16] present an IoT gateway system based on the GPRS and Zigbee protocols. They also explain that the applications of the IoT in telecommunications networks, to control the functions of sensor networks. Rathee et al. [17] propose a new model called the fog computing environment for IoT devices. They study a secure routed and handoff mechanism to minimize attacks by exploring the true value and measure of each and every fog IoT device, based on their communication behavior. K. Angrishi [18] explains the IoTs are affected through botnets, on the Internet connection. He also analyzes how botnets affect the IoT. B. K. Mishra et al.[19] present a mathematical model on the DDoS attack through the IoTs in a network. They mainly focus on the impact of internal and external nodes attack on IoT devices, based on the DDoS attack. Yerra Shankar Rao et al. [20] show the DDoS attack on the target resources in a network for the critical infrastructure. They mainly focus on the target resource and apply quarantine node-making for defense. Fan Dang et al. [21] explain a file-less attack on a Linux-based IoT with the honey cloud. They deploy the hardware and software honeypot and measure the impact of a fileless real-world attack on the IoT. IoT device attacks are mainly based on malware that can be transmitted quickly in IoT devices. They analyze the static and dynamic models to restrict malware attacks. However, these explanations are found to be inadequate as an overall explanation for the phenomenon of attacks by malware in IoT-device systems. In this chapter, the authors use the technique of the vaccinated model (SVEIR) to develop and establish a mathematical relationship to control malware attacks in IoT devices. Malware behaves like infectious diseases and epidemics in nature. Mathematical models generalize to represent the behavior of many kinds of malware in IoT devices. An epidemical model explains the propagation of malware and its agents in the closed system considered by Kermack and McKendrick [22]. This theory is applied biologically as well as to its technological counterpart. Vaccination is an important concept of the technical process that eliminates/controls the infectious malware (Mirai) in a network connected by the Internet.

This chapter is organized into the following sections. Section 7.1 introduces the chapter. Section 7.2 presents the formulation of the e-epidemic (mathematical) model. Section 7.3 analyzes the basic reproduction number. Section 7.4 focuses on stability analysis. Section 7.5 reveals the numerical and simulation of the model. The conclusion of the model is discussed in Section 7.6.

7.1.1 Basic Terminology

N: Total number of nodes interacting with each other in the IoT devices
S: Susceptible nodes interacting with IoT devices in the network
E: Exposed nodes becoming infected for a short period of time
I: Infectious nodes infected by interacting with malware
R: Recovered nodes
R_0: Basic reproduction number
A: Rate of newborn nodes entering the network
β: Infectivity contact rate
α: Infectious rate
γ: The rate at which infectious nodes recover
δ: The rate at which susceptible nodes are vaccinated
ε: The rate at which vaccinated nodes become infectious due lack of anti-malware software
η: The rate at which recovered nodes become susceptible
μ_1: Natural death
μ_2: Death tare other than malware attack

7.2 HYPOTHESIS AND MATHEMATICAL MODEL FORMULATION

The subsequent postulation/hypothesis are taken into consideration to describe the mathematical model.

1. All attacking nodes are malware-free and either susceptible or infectious due to disconnection from the Internet.
2. The attacking node undergoes a latent period (exposed nodes), then, after infection, is converted into an infectious node.
3. Once the Internet is switched on, the attacking node becomes infected by contact with infected nodes at the rate βSI, where β is the positive constant.
4. Because our model is dynamic, each and every attacking node dies naturally at the rate μ_1.
5. The attacking node crashes due to network failure or lack of an Internet connection at the rate μ_2.
6. Exposed nodes become infectious at the rate α.

7. Susceptible nodes become vaccinated after treatment by a primary scan at the rate δ.
8. Vaccinated nodes become infected due a lack of updated anti-malware software at the rate ε.
9. Due to temporary immunity, the recovered nodes become susceptible at the rate η.
10. Infectious nodes recover after using anti-malware software at the rate γ.

7.2.1 Mathematical Model Formulation

Based upon these assumptions, we developed an e-epidemic model. In this model, we assume that the total number of nodes consists of five compartments of nodes: susceptible, exposed, vaccinated, infected, and recovery nodes. Because of this, we will take a dynamic model the flow of botnets (Mirai) is from S to E, S to V, V to I, E to I, I to R, and R to S compartments, as shown in Figure 7.1. The propagation of malware (Mirai) can be represented by a system of ordinary differential equations.

Mathematical equations

$$\frac{dS}{dt} = A - \beta SI - (\mu_1 + \delta)S + \eta R$$
$$\frac{dV}{dt} = \delta S - (\mu_1 + \mu_2 + \varepsilon)V$$
$$\frac{dE}{dt} = \beta SI - (\mu_1 + \alpha)E \qquad (7.1)$$
$$\frac{dI}{dt} = \alpha E - (\mu_1 + \mu_2 + \gamma)I + \varepsilon V$$
$$\frac{dR}{dt} = \gamma I - (\mu_1 + \eta)R$$

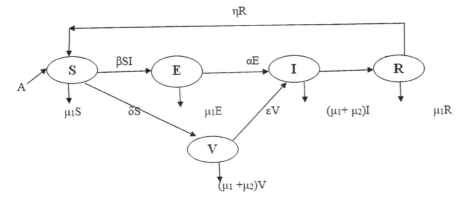

FIGURE 7.1 Diagrammatical representation of the model.

7.2.2 MATHEMATICAL MODEL ANALYSIS

Mathematical analysis can provide a theoretical foundation for predicting malware transmission in IoT devices. In this section, we assume that the initial conditions of the system (1) are non-negative.

It is important to show the positivity and boundedness of the system of the foregoing equations as they are represented by the different nodes in the network. So we can add all the above equations

$$\frac{dN}{dt} = \frac{dS}{dt} + \frac{dV}{dt} + \frac{dE}{dt} + \frac{dI}{dt} + \frac{dR}{dt}$$

$$\Rightarrow \frac{dN}{dt} = A - \mu_1 N - \mu_2 (I + V)$$

Due to the absence of any malware in the IoT devices, ($I=V=0$), it follows that $\frac{dN}{dt} = A - \mu_1 N$.

$$\Omega = \left\{ \begin{array}{l} S(t), V(t), E(t), I(t), R(t) \in R_+^5 : S(0) \ge 0, V(0) \ge 0, E(0) \ge 0, I(0) \ge 0, R(0) \ge 0, \\ S(t) + V(t) + E(t) + I(t) + R(t) \le \dfrac{A}{\mu_1} \end{array} \right\}$$

Thus, this feasible region Ω is positively invariant with respect to system (1), when for all $t \ge 0, as, t \to \infty, then\ N(t) \to \dfrac{A}{\mu_1}$. Therefore, every solution of the system (1) with initial condition in the region remains positive. Hence, our model is mathematically well posed.

7.3 EQUILIBRIUM POINTS AND BASIC REPRODUCTION NUMBER

The crash/success for any type of malware attack on IoT-device signals depends on the basic reproduction number. During the entire life cycle of the network system, the basic reproduction number is defined as the average number of secondary infections caused in a totally susceptible node by single infectious nodes. This basic reproduction number plays an important role for both biological as well as technical aspects. In technical terms, it determines whether the malware will die or persist in the network

Mathematically, it can be calculated by linearizing the system of equations. We get

$$\begin{bmatrix} \dfrac{dE}{dt} \\ \dfrac{dI}{dt} \\ \dfrac{dV}{dt} \end{bmatrix} = (F - K) \begin{bmatrix} E \\ I \\ V \end{bmatrix}$$

Transmission Modeling on Malware Attack

where F is the matrix of rates of infections and K the matrix of transmission rates from compartment to compartment, defined as

$$F = \begin{pmatrix} 0 & \beta & 0 \\ 0 & 0 & 0 \\ 0 & 0 & 0 \end{pmatrix}, K = \begin{pmatrix} \left(\mu_1 + \alpha\right) & 0 & 0 \\ -\alpha & \left(\mu_1 + \mu_2 + \gamma\right) & -\varepsilon \\ 0 & 0 & \left(\mu_1 + \mu_2 + \varepsilon\right) \end{pmatrix}$$

Then, the basic reproduction number R_0 is defined as the dominant eigen values of FK^{-1}.

$$R_0 = \rho\left(FK^{-1}\right) = \frac{\beta\alpha\varepsilon}{\left(\mu_1 + \mu_2 + \gamma\right)\left(\mu_1 + \alpha\right)}$$

7.3.1 EXISTENCE OF STABILITY OF THE EQUILIBRIUM POINTS

In this section, we discuss two types of equilibrium points forthe system and investigate the system's dynamic nature.

1. Infection free (malware-free equilibrium point)
2. Endemic equilibrium points

For the malware-free equilibrium point, in the absence of any type of IoT-based malware attack, the network will remain $\Omega_0 = (I = 0, E = 0, V = 0, S = 1)$.

In the case of an endemic equilibrium point, the infection remains in the IoT devices. That is, the network of the system becomes $\Omega^* = (S^*, E^*, I^*, V^*, R^*)$

$$S^* = \frac{A\left(\mu_1 + \eta\right) + \eta\gamma I^*}{\left(\mu_1 + \eta\right)\left(\mu_1 + \delta + \beta I^*\right)}$$

$$V^* = \frac{\delta A\left(\mu_1 + \eta\right) + \delta\eta\gamma I^*}{\left(\mu_1 + \eta\right)\left(\mu_1 + \mu_2 + \varepsilon\right)\left(\mu_1 + \delta + \beta I^*\right)}$$

$$E^* = \frac{\beta I^*\left[A\left(\mu_1 + \eta\right) + \eta\gamma I^*\right]}{\left(\mu_1 + \eta\right)\left(\mu_1 + \delta + \beta I^*\right)\left(\alpha + \mu_1\right)}$$

$$R^* = \frac{\gamma I^*}{\left(\mu_1 + \eta\right)}$$

7.3.2 LOCAL STABILITY OF THE MALWARE-FREE EQUILIBRIUM POINT

Theorem 7.1

If $R_0 \leq 1$, then the malware-free equilibrium point $\Omega_0 = (I = 0, E = 0, V = 0, S = 1)$ of system (1) is locally asymptotical stable in the given region and it unstable when $R_0 > 1$.

Proof: Linearized, the malware-free equilibrium point $\Omega_0 = (I = 0, E = 0, V = 0, S = 1)$ of system (1) becomes, in the Jacobian matrix,

$$J_{MFE} = \begin{pmatrix} -(\mu_1 + \delta) & 0 & -\beta & 0 \\ 0 & -(\alpha + \mu_1) & \beta & 0 \\ 0 & \alpha & -(\mu_1 + \mu_2 + \gamma) & \varepsilon \\ \delta & 0 & 0 & -(\mu_1 + \mu_2 + \varepsilon) \end{pmatrix}$$

The characteristic equation of the foregoing Jacobian matrix is

$$\lambda^4 + a_1\lambda^3 + a_2\lambda^2 + a_3\lambda + a_4 = 0$$

Where

$$a_1 = \left(4\mu_1 + \delta + 2\mu_2 + \gamma + \mu_1 + \mu_2 + \varepsilon + \alpha \right)$$

$$a_2 = \begin{bmatrix} (\mu_1 + \delta)(\alpha + \mu_1) + (\alpha + \mu_1)(\mu_1 + \mu_2 + \varepsilon) + (\mu_1 + \mu_2 + \varepsilon)(\mu_1 + \delta) \\ + (\mu_1 + \delta)(\mu_1 + \mu_2 + \gamma) + (\mu_1 + \mu_2 + \gamma)(\mu_1 + \mu_2 + \varepsilon) + (\alpha + \mu_1)(\mu_1 + \mu_2 + \gamma) - \beta\alpha \end{bmatrix}$$

$$a_3 = \begin{bmatrix} (\mu_1 + \delta)(\alpha + \mu_1)(\mu_1 + \mu_2 + \varepsilon) + (\mu_1 + \mu_2 + \varepsilon)(\mu_1 + \delta)(\mu_1 + \mu_2 + \gamma) \\ + (\mu_1 + \mu_2 + \gamma)(\mu_1 + \delta)(\alpha + \mu_1) + (\mu_1 + \mu_2 + \gamma)(\mu_1 + \mu_2 + \varepsilon)(\alpha + \mu_1) \\ -(\mu_1 + \delta)\beta\alpha - \alpha\beta(\mu_1 + \mu_2 + \varepsilon) + \beta\delta\varepsilon \end{bmatrix}$$

$$a_4 = \begin{bmatrix} (\mu_1 + \delta)(\alpha + v\mu_1)(\mu_1 + \mu_2 + \varepsilon)(\mu_1 + \mu_2 + \gamma) + \beta\delta\varepsilon(\alpha + \mu_1) - (\mu_1 + \delta)(\mu_1 + \mu_2 + \varepsilon)\alpha\beta \end{bmatrix}$$

Since all a_1, a_2, a_3, a_4 are positive. So,

$$a_1 > 0, a_3 > 0, a_4 > 0$$
$$And$$
$$a_1a_2a_3 > a_3^2 + a_1^2a_4$$
$$\Rightarrow a_1a_2a_3 - \left(a_3^2 + a_1^2a_4 \right) > 0$$

Hence, by Rauth-Hurwitz criteria for the polynomial, system (1) at the malware-free equilibrium is locally asymptotically stable in the given region.

7.3.3 LOCAL STABILITY OF THE ENDEMIC EQUILIBRIUM

Theorem 7.2

When $R_0 > 1$, there exists a unique endemic equilibrium in the region. $\Omega^* = (S^*, E^*, I^*, V^*, R^*)$ is locally asymptotically stable in the preceding region.

Transmission Modeling on Malware Attack 111

Proof: At the endemic equilibrium point $\Omega^* = (S^*, E^*, I^*, V^*, R^*)$ of system (1), the Jacobian matrix is

$$J_{EE} = \begin{pmatrix} -\left(\mu_1 + \delta + \beta I^*\right) & 0 & -\beta S^* & 0 & \eta \\ \beta I^* & -\left(\alpha + \mu_1\right) & \beta S^* & 0 & 0 \\ 0 & \alpha & -\left(\mu_1 + \mu_2 + \gamma\right) & \varepsilon & 0 \\ \delta & 0 & 0 & -\left(\mu_1 + \mu_2 + \varepsilon\right) & 0 \\ 0 & 0 & \gamma & 0 & -\left(\mu_1 + \eta\right) \end{pmatrix}$$

The characteristic of the root of the equation is

$$\lambda^5 + b_1\lambda^4 + b_2\lambda^3 + b_3\lambda^2 + b_4\lambda + b_5 = 0$$

Where

$$b_1 = \left[5\mu_1 + \delta + \beta I^* + \alpha + 2\mu_2 + \gamma + \varepsilon + \eta\right]$$

$$b_2 = \begin{bmatrix} \left(\mu_1 + \delta + \beta I^*\right)\left(\alpha + \mu_1\right) + \left(\mu_1 + \delta + \beta I^*\right)\left(\mu_1 + \mu_2 + \gamma\right) + \left(\mu_1 + \delta + \beta I^*\right)\left(\mu_1 + \mu_2 + \varepsilon\right) \\ +\left(\mu_1 + \delta + \beta I^*\right)\left(\mu_1 + \eta\right) + \left(\alpha + \mu_1\right)\left(\mu_1 + \mu_2 + \gamma\right) + \left(\mu_1 + \mu_2 + \varepsilon\right)\left(\alpha + \mu_1\right) + \\ \left(\alpha + \mu_1\right)\left(\mu_1 + \eta\right) + \left(\mu_1 + \eta\right)\left(\mu_1 + \mu_2 + \gamma\right) + \left(\mu_1 + \mu_2 + \gamma\right)\left(\mu_1 + \mu_2 + \varepsilon\right) + \\ \left(\mu_1 + \mu_2 + \varepsilon\right)\left(\mu_1 + \eta\right) - \beta S^*\alpha \end{bmatrix}$$

$$b_3 = \begin{bmatrix} \left(\mu_1 + \delta + \beta I^*\right)\left(\alpha + \mu_1\right)\left(\mu_1 + \mu_2 + \gamma\right) + \left(\mu_1 + \delta + \beta I^*\right)\left(\alpha + \mu_1\right)\left(\mu_1 + \mu_2 + \varepsilon\right) + \\ \left(\mu_1 + \delta + \beta I^*\right)\left(\alpha + \mu_1\right)\left(\mu_1 + \eta\right) + \left(\mu_1 + \delta + \beta I^*\right)\left(\mu_1 + \mu_2 + \gamma\right)\left(\mu_1 + \mu_2 + \varepsilon\right) + \\ \left(\mu_1 + \eta\right)\left(\mu_1 + \delta + \beta I^*\right)\left(\mu_1 + \mu_2 + \gamma\right) + \left(\mu_1 + \eta\right)\left(\mu_1 + \delta + \beta I^*\right)\left(\mu_1 + \mu_2 + \varepsilon\right) + \\ \left(\alpha + \mu_1\right)\left(\mu_1 + \mu_2 + \gamma\right)\left(\mu_1 + \mu_2 + \varepsilon\right) + \left(\alpha + \mu_1\right)\left(\mu_1 + \mu_2 + \gamma\right)\left(\mu_1 + \eta\right) + \\ \left(\alpha + \mu_1\right)\left(\mu_1 + \mu_2 + \varepsilon\right)\left(\mu_1 + \eta\right) + \left(\left(\mu_1 + \mu_2 + \gamma\right)\left(\mu_1 + \mu_2 + \gamma\right)\left(\mu_1 + \eta\right) + \beta S^*\beta I^*\alpha \\ +\beta S^*\varepsilon\delta + \left(\mu_1 + \delta + \beta I^*\right)\alpha\beta S^* + \alpha\beta S^*\left(\mu_1 + \mu_2 + \varepsilon\right) + \alpha\beta S^*\left(\mu_1 + \eta\right) \end{bmatrix}$$

$$b_4 = \begin{bmatrix} \left(\mu_1 + \delta + \beta I^*\right)\left(\alpha + \mu_1\right)\left(\mu_1 + \mu_2 + \gamma\right)\left(\mu_1 + \mu_2 + \varepsilon\right) + \left(\mu_1 + \delta + \beta I^*\right)\left(\alpha + \mu_1\right)\left(\mu_1 + \mu_2 + \gamma\right)\left(\mu_1 + \eta\right) + \\ \left(\mu_1 + \eta\right)\left(\mu_1 + \delta + \beta I^*\right)\left(\alpha + \mu_1\right)\left(\mu_1 + \mu_2 + \varepsilon\right) + \left(\mu_1 + \eta\right)\left(\mu_1 + \mu_2 + \varepsilon\right)\left(\mu_1 + \mu_2 + \gamma\right)\left(\mu_1 + \delta + \beta I^*\right) + \\ \left(\mu_1 + \eta\right)\left(\mu_1 + \mu_2 + \varepsilon\right)\left(\mu_1 + \mu_2 + \gamma\right)\left(\alpha + \mu_1\right) + \beta I^*\beta S^*\alpha\left(\mu_1 + \mu_2 + \varepsilon\right) + \beta I^*\beta S^*\alpha\left(\mu_1 + \eta\right) + \\ \beta S^*\left(\alpha + \mu_1\right)\delta\varepsilon + \beta S^*\left(\eta + \mu_1\right)\delta\varepsilon = \eta\alpha\gamma\beta I^* - \eta\delta\varepsilon\gamma - \left(\mu_1 + \delta + \beta I^*\right)\beta S^*\alpha\left(\mu_1 + \mu_2 + \varepsilon\right) \\ -\left(\mu_1 + \delta + \beta I^*\right)\beta S^*\alpha\left(\mu_1 + \eta\right) - \left(\mu_1 + \mu_2 + \varepsilon\right)\beta S^*\alpha\left(\mu_1 + \eta\right) \end{bmatrix}$$

$$b_5 = \begin{bmatrix} \left(\mu_1 + \delta + \beta I^*\right)\left(\alpha + \mu_1\right)\left(\mu_1 + \mu_2 + \gamma\right)\left(\mu_1 + \mu_2 + \varepsilon\right)\left(\mu_1 + \eta\right) + \beta S^* \beta I^* \alpha \left(\mu_1 + \mu_2 + \varepsilon\right)\left(\mu_1 + \eta\right) + \\ \left(\alpha + \mu_1\right)\left(\mu_1 + \eta\right)\beta S^* \varepsilon \delta + \left(\mu_1 + \delta + \beta I^*\right)\left(\mu_1 + \mu_2 + \varepsilon\right)\left(\mu_1 + \eta\right)\beta S^* \alpha - \left(\mu_1 + \mu_2 + \varepsilon\right)\beta I^* \eta \alpha \gamma - \left(\alpha + \mu_1\right)\eta \delta \varepsilon \gamma \end{bmatrix}$$

Since

$$b_1 b_2 b_3 > b_3^2 + b_1^2 b_4$$

&

$$\left[\left(b_1 b_4 - b_5\right)\left(b_1 b_2 b_3 - \left(b_3^2 + b_1^2 b_4\right)\right)\right] - \left[b_5\left(b_1 b_2 - b_3\right)^2 + b_1 b_5^2\right] > 0$$

So, by the Routh-Hurwitz Condition for stability, the system is asymptotically locally stable at endemic equilibrium point.

7.4 GLOBAL STABILITY OF THE EQUILIBRIUM POINTS

Theorem 7.3

The malware-free equilibrium point is globally asymptotically stable if $R_0 \leq 1$.

Proof: Consider the Lyaponov function

$$L = \alpha E + \varepsilon I$$

$$\Rightarrow \frac{dL}{dt} = \alpha \frac{dE}{dt} + \varepsilon \frac{dI}{dt}$$

$$\Rightarrow \frac{dL}{dt} = \alpha \left[\beta SI - \left(\mu_1 + \alpha\right)E\right] + \varepsilon \left[\alpha E - \left(\mu_1 + \mu_2 + \gamma\right)I + \varepsilon V\right]$$

$$\Rightarrow \frac{dL}{dt} = I\alpha \left[\beta S - \left(\mu_1 + \mu_2 + \gamma\right)\right] - \left(\mu_1 + \alpha\right)E + \alpha E + \varepsilon V$$

$$\Rightarrow \frac{dL}{dt} = I\alpha \left[\frac{\beta S}{\left(\mu_1 + \mu_2 + \gamma\right)} - 1\right]\left(\mu_1 + \mu_2 + \gamma\right) - \left(\mu_1 + \alpha\right)E + \alpha E + \varepsilon V$$

$$\Rightarrow \frac{dL}{dt} = I\left[\frac{\beta S\alpha}{\left(\mu_1 + \alpha\right)\left(\mu_1 + \mu_2 + \gamma\right)} - 1\right]\left(\mu_1 + \alpha\right)\left(\mu_1 + \mu_2 + \gamma\right) - E + \frac{\alpha E + \varepsilon V}{\left(\mu_1 + \alpha\right)}$$

$$\Rightarrow \frac{dL}{dt} = I\left[R_0 - 1\right]\left(\mu_1 + \alpha\right)\left(\mu_1 + \mu_2 + \gamma\right) - E + \frac{\alpha E + \varepsilon V}{\left(\mu_1 + \alpha\right)} \leq 0$$

$$\Rightarrow \frac{dL}{dt} \leq 0$$

if $R_0 \leq 1 \Rightarrow \dfrac{dL}{dt} \leq 0$ and $\dfrac{dL}{dt} = 0 \Rightarrow R_0 = 1$

Hence, by Lassalle's maximum invariant principle, the malware-free equilibrium point is globally stable in the given region of the IoT devices.

Transmission Modeling on Malware Attack

113

7.4.1 GLOBAL STABILITY FOR ENDEMIC EQUILIBRIUM POINT

Theorem 7.4

When $R_0 > 1$, then the endemic equilibrium is asymptotically globally stable in the region Ω^*.

Proof: Let's consider the nonlinear Goh-Volterra type Lyapunov function

$$Z = \left(S - S^* - S^* \ln\frac{S}{S^*}\right) + \left(V - V^* - V^* \ln\frac{V}{V^*}\right) + \left(E - E^* - E^* \ln\frac{E}{E^*}\right) + \frac{T_1}{\alpha}\left(I - I^* - I^* \ln\frac{I}{I^*}\right)$$

$$\Rightarrow \frac{dZ}{dt} = \left[\frac{dS}{dt} - S^*\frac{\frac{dS}{dt}}{S}\right] + \left[\frac{dV}{dt} - V^*\frac{\frac{dV}{dt}}{V}\right] + \left[\frac{dE}{dt} - E^*\frac{\frac{dE}{dt}}{E}\right] + \frac{T_1}{\alpha}\left[\frac{dI}{dt} - I^*\frac{\frac{dI}{dt}}{I}\right]$$

$$\Rightarrow \frac{dZ}{dt} = \left\{ \begin{array}{l} \left[A - \beta SI - (\mu_1 + \delta)S + \eta R\right] - \frac{S^*}{S}\left[A - \beta SI - (\mu_1 + \delta)S + \eta R\right] + \\[6pt] \left[\delta S - (\mu_1 + \mu_2 + \varepsilon)V\right] - \frac{V^*}{V}\left[\delta S - (\mu_1 + \mu_2 + \varepsilon)V\right] + \left[\beta SI - (\mu_1 + \alpha)E\right] - \\[6pt] \frac{E^*}{E}\left[\beta SI - (\mu_1 + \alpha)E\right] + \frac{T_1}{\alpha}\left[\alpha E - (\mu_1 + \mu_2 + \gamma)I + \varepsilon V\right] - \frac{T_1}{\alpha}\frac{I^*}{I}\left[\alpha E - (\mu_1 + \mu_2 + \gamma)I + \varepsilon V\right] \end{array} \right\}$$

$$\Rightarrow \frac{dZ}{dt} = \left\{ \begin{array}{l} \left[A - \beta SI - (\mu_1 + \delta)S + \eta R\right] - \frac{S^*}{S}\left[A - \beta SI - (\mu_1 + \delta)S + \eta R\right] + \\[6pt] \left[\delta S - (\mu_1 + \mu_2 + \varepsilon)V\right] - \frac{V^*}{V}\left[\delta S - (\mu_1 + \mu_2 + \varepsilon)V\right] + \left[\beta SI - (\mu_1 + \alpha)E\right] - \\[6pt] \frac{E^*}{E}\left[\beta SI - (\mu_1 + \alpha)E\right] + \frac{T_1}{\alpha}\left[\alpha E - T_2 I + \varepsilon V\right] - \frac{T_1}{\alpha}\frac{I^*}{I}\left[\alpha E - T_2 I + \varepsilon V\right] \end{array} \right\}$$

Here, we can take $T_1 = (\mu_1 + \alpha)$, $T_2 = (\mu_1 + \mu_2 + \gamma)$ & $T_3 = (\mu_1 + \delta)$
From the steady state condition of the system of the equation, we have

$$A = \beta S^* I^* - T_3 S^* + \eta R^*$$

$$T_1 = \frac{\beta S^* I^*}{E^*},$$

$$T_2 = \frac{\alpha E^* - \varepsilon V^*}{I^*},$$

$$T_3 = (\mu_1 + \delta$$

Using the preceding values of A, T_1, T_2, T_3, then adding and simultaneously subtracting $\dfrac{\beta I^* V^{2^*}}{V}$ & $\beta V^* I^*$, we will find the results

$$\frac{dZ}{dt} = \left\{ \begin{array}{l} T_3 S^*\left[2 - \frac{S}{S^*} - \frac{S^*}{S}\right] + \beta S^* I^*\left[3 - \frac{S^*}{S} - \frac{SE^* I}{S^* EI^*} - \frac{EI^*}{IE^*}\right] + \\[8pt] \beta V^* I^*\left[3 - \frac{VE^* I}{V^* I^* E} - \frac{EI^*}{IE^*} - \frac{V^*}{V}\right] - \left[\frac{V - V^*}{V}\right]\left[\delta S - \varepsilon V + \beta V^* I^*\right] \end{array} \right\}$$

114 Cybersecurity

Since the arithmetic mean is more than the geometric mean, then the result is good for the inequality. So,

$$\left[2 - \frac{S}{S^*} - \frac{S^*}{S}\right] \leq 0,$$

$$\left[3 - \frac{S^*}{S} - \frac{SE^*I}{S^*EI^*} - \frac{EI^*}{IE^*}\right] \leq 0,$$

$$\left[3 - \frac{VE^*I}{V^*I^*E} - \frac{EI^*}{IE^*} - \frac{V^*}{V}\right] \leq 0$$

At the endemic state $\frac{dV}{dt} = 0$, so the last term of the preceding expression is zero. Hence, $\frac{dZ}{dt} \leq 0$.

Thus, Z satisfies LaSalle's invariant principle [23], which gives the solutions of the result proven by the LaSalle principle, where the initial conditions have a unique endemic equilibrium that is globally stable in the region when the basic reproduction number is greater than one.

Therefore, the endemic equilibrium point is globally asymptotically stable in the region Ω^*.

7.5 NUMERICAL SIMULATION AND RESULTS

In this section, we provide a numerical simulation carried by MATLAB, supporting the theoretical analysis done in the preceding section. Figure 7.2 represents the malware-free equilibrium by choosing some parameter under $R_0 < 1$. In this case, the susceptible node goes to a steady state condition, while the exposed infectious, vaccinated nodes and recovered nodes go to zero as time goes to infinity. That means the malware (Mirai) attack will disappear and dies out in the IoT-device network. Figure 7.3 explains the endemic equilibrium point when $R_0 > 1$. Here, all the nodes approach a steady state condition, and time goes to infinity. As a result, the malware will remain in the IoT network, and the network becomes endemic in condition. Thus, the vaccination process is an effective method for reducing infectious nodes in the network, so that they do not spread an infection of Mirai (malware) in the IoT devices. If we increase the vaccination rate δ, the infective rate will decrease, as shown in Figure 7.4. Also, the vaccinated nodes becomes infected, due to a lack of anti-malware with the latest signature. Similarly, if we increase the vaccination rate, then the rate of recovery increases simultaneously, as depicted in Figure 7.5. Hence, this indicates that the spread of malware decreases as protective measures such as vaccination (anti-malware) for the susceptible nodes increase in IoT-device networks. This type of protection can be applied to websites, which can be filter software and prevent unwanted malware in IoT devices.

Transmission Modeling on Malware Attack

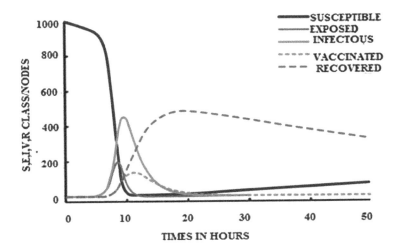

FIGURE 7.2 Dynamic behavior of the model when $R_0 < 1$.

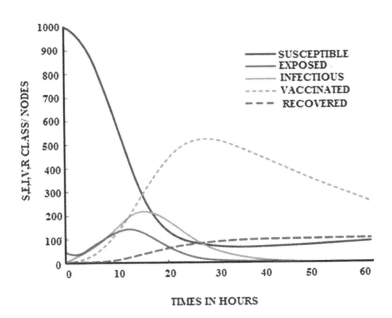

FIGURE 7.3 Dynamic behavior of the nodes when $R_0 > 1$.

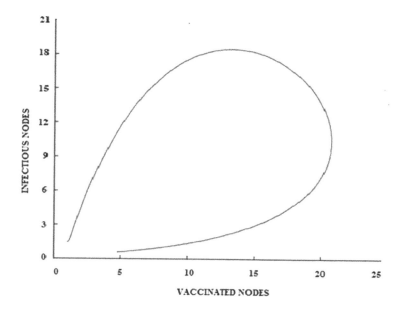

FIGURE 7.4 Effect of infected nodes versus vaccinated nodes.

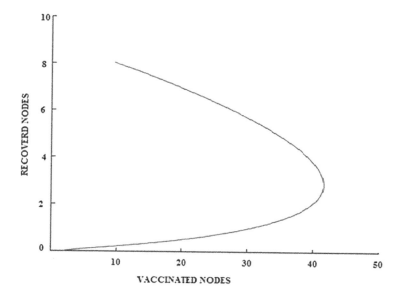

FIGURE 7.5 Effect of vaccinated nodes versus recovered nodes.

7.6 CONCLUSION

In this chapter, we propose an e epidemical model and analyze it analytically as well as numerically for the propagation of botnet (Mirai) attacks through IoT devices. In order to fit the botnet outbreak in the network, a vaccination-like cellular firewall is introduced from the beginning to block the attack in the network. Our model is based on a bonnet attack, where the attacker becomes apparent in the network through IoT devices. The success or failure of the attack depends on the basic reproduction number. If this number is more than one, the malware attack will spread, but below unity, the malware attack will disappear or die in IoT devices. The impact of vaccination is very important to block the attack from the devices in the network. This model suggests that upgrading the fireware of smart devices regularly, reliable authentication mechanisms, and blocking the process at malware entry points, could control the attack from the malware. Thus, IoT devices would be malware-free, if we updated the anti-malware from time to time. Therefore, our model is helpful for software companies or organizations to protect IoT-device networks from attack. Our future work will be extended to scale free network and time-delay model parameters for encryption or decryption of the IoT devices in the networks.

REFERENCES

[1] M. T. Gardner, C. Beard, D. Medhi. 2017. *"Using SEIRS epidemic models for IoT botnets attacks."* In *Proceeding of 13th International Conferenceon Design of Reliable Communication Networks*, Munich, Germany, March 08-10, 2017, pp. 62–69. Berlin: VDE.

[2] Y. Ji, L. Yao, S. Liu, H. Yao, Q. Ye, R. Wang. 2018, 9–11 May. *"The study on the botnet and its prevention policies in the IoTs."* In *Proceedings of 22nd International Conference on Computer Supported Cooperative Work in Design (CSCWD)*, pp. 837–842. Nanjing, China: IEEE.

[3] D. Acarali, M. Rajarajan, N. Komninos, B. B. Zarpelão. 2019. "Modelling the spread of botnet malware in IoT-based wireless sensor networks." *Security and Communication Networks*, 2019, pp. 1–13

[4] P. Hejazi, G. Ferrari. 2018. "Energy and memory efficient data loss prevention in wireless sensor networks." *Sensors*, pp. 1–18. Preprints 2018, 2018070206 (doi: 10.20944/ preprints201807.0206.v1).

[5] C. Kolias, G. Kambourakis, A. Stavrou, J. Voas. 2017. "DDoS in the IoT: Mirai and other botnets" *IEEE Computer Society*, 50(7), pp. 80–84.

[6] Y. S. Rao, H. Saini, G. T. C. Panda. 2019, April 12–13. *"Effect of vaccination in the computer network for distributed attacks: A dynamic model."* *Advances in Computing and Data Sciences Third International Conference, ICACDS 2019*, Ghaziabad, India, pp. 175–184. Switzerland AG: Springer Nature. ISBN: 978-981-13-9941-1

[7] H. Akram, H. A. Abdul-Ghani, D. Konstantas. 2019. "A comprehensive study of security and privacy guidelines, threats, and countermeasures: An IoT perspective" *Journal of Sensor and Actuator Networks*, 8(2), p. 38. DOI: 10.3390/jsan8020022

[8] Y. S. Rao, H. Saini, R. Rath, T. C. Panda. 2020. "Dynamic modeling on malware and its defense in wireless computer network using pre-quarantine." In Gautam Kumar, Dinesh Kuman Saini, Nguyen Ha Huy Cuong (Eds.), *Cyber Defense Mechanisms*. Boca Raton: CRC Press (Taylor & Francis). ISBN 9780367408831 2020/9/20, pp. 171–183.

[9] C. Zhang, R. Green. 2015. *"Communication security in IoT: Preventive measure and avoid DDoS attack over IoT network."* In *Proceedings of the 18th Symposium on Communications & Networking. Society for Computer Simulation International*, pp. 8–15. Alexandria, VA: ACM.

[10] M. J. Farooq, Q. Zhu. 2019. "Modelling, analysis, and mitigation of dynamic botnet formation in wireless IoT networks." *IEEE Transactions on Information Forensics and Security*, 14(9), pp. 2412–2426.

[11] M. Sudip, P. Venkata Krishna, H. Agarwal, A. Saxena, M. S. Obaidat. 2011. *"A learning automata based solution for preventing distributed denial of service in IoTs."* In *International Conferences on IoTs, Cyber, Physical and Social Computing*, pp. 114–122. Dalian, China: IEEE.

[12] Z. Daqiang, L. T. Yang, H. Huang. 2011. *"Searching in IoTs: Vision and challenges."* In *International Symposium on Parallel and Distributed Processing with Applications*, pp. 206–211. Los Alamitos, CA: IEEE Computer Society. doi:10.1109/ISPA.2011.

[13] R. Sfar, E. Natalizio, Y. Challal, Z. Chtourou. 2018. "A roadmap for security challenges in the IoTs."*Digital Communications and Networks*, 4, pp. 118–137.

[14] J. A. Jerkins, J. Stupiansky. 2018. *"Mitigating IoT insecurity with inoculation epidemics."* In *Proceedings of the ACMSE 2018 Conference*, p. 4. Richmond, KY: ACM.

[15] M. M. R. Singh, A. Gahlot, R. Samkaria.2019. "IoT based intelligent robot for various disasters monitoring and prevention with visual data manipulating." *International Journal of Tomography and Simulation*, 32(1), pp. 90–99.

[16] Q. Zhu, R. Wang, Q. Chen, Y. Liu, W. Qin. 2010.*"IoT gateway: Bridging wireless sensor networks into IoTs."* In *2010 IEEE/IFIP International Conference on Embedded and Ubiquitous Computing*, pp. 347–352. Hong Kong, China: IEEE. https://doi.org/10.1109/EUC.2010.58

[17] G. Rathee, R. Sandhu, H. Saini, M. Sivaram, V. Dhasaratha. 2020. "A trust computed framework for Iot devices and fog computing environments." *WirelessNetworks*, 26(4), pp. 2339–2351.

[18] K. Angrishi. 2017. "Turning Internet of Things (IoT) into Internet of Vulnerabilities (IoV): IoT botnets." *arXiv*, 2017, arXiv: 1702.0368

[19] B. K. Mishra, A. K. Keshri, D. K. Mallick, B. K. Mishra. 2018. "Mathematical model on distributed denial of service attack through Internet of Things in a network." *Nonlinear Engineering*, 8(1), pp. 486–495. https://doi.org/10.1515/nleng-2017-0094.

[20] Y. S. Rao, A. Keshri, B. K. Mishra, T. C. Panda. 2019. "Distributed denial of service attack on targeted resources in a computer network for critical infrastructure: A differential e-epidemic model." *Physica A: Statistical Mechanics and its Applications*, 540, 123240. https://doi.org/10.1016/j.physa.2019.123240

[21] F. Dang, Z. Li, Y. Liu, E. Zhai, Q. A. Chen, T. Xu, Y. Chen, J. Yang. 2019, June 17–21. *"Understanding fileless attacks on linux-based IoT devices with honey cloud."* In *MobiSys '19*, Session 9: Nuts and Bolts, pp. 482–493. Seoul, Korea: ACM.

[22] W. O. Kermack, A. G. McKendrick. 1933. *"Contribution of mathematical theory to epidemics."*Proceedings of the Royal Society of London, Series A*, 141, pp. 94–122.

[23] J. P. LaSalle. 1976. *"The stability of dynamical systems."* In *CBMS-NSF Regional Conference Series in Applied Mathematics*, pp. 418–420. Philadelphia: SIAM.

8 Rice Plant Disease Detection Using IoT

FarjanaYeasmin Trisha
East West University, Bangladesh

Mahmudul Hasan
Jahangirnagar University, Bangladesh

CONTENTS

8.1 Introduction 119
8.2 Related Work 122
8.3 Proposed System Model 123
    8.3.1 Flowchart of the Following System 124
8.4 Circuit Diagram 127
8.5 Result 128
8.6 Conclusion 129
References 129

8.1 INTRODUCTION

Rice is a mainly consumed food for most of the population, a natural source of carbohydrates, protein, and dietary fiber, and one of the most developed foods all over the world. It is frequently used as a part of daily life. Rice is the most generally powerful supplement food accessible in Asia [1]. Rice is cultivated in five regions, including Asia, America, Africa, Europe, and Australia. On a global scale, Bangladesh is third in rice cultivation [2]. Rice production there is a very important part of the national economy. Bangladesh achieved its highest GDP, BDT 10.73 billion in 2019, from the agricultural sector [3]. Rice production is one of the most important of all agricultural sectors. Half of the agricultural GDP is provided by rice production. While making an essential part of the nation's economy, rice fills in as staple nourishment for the mass populace and gives 66% of the per capita daily calorie intake [4]. Figure 8.1 Shew et al. (2019) shows the paddy production rate of the country.

Disease-free rice production would assume a prevailing part in guaranteeing stable monetary development and keeping up the ideal financial targets. The population is increasing continuously day by day, so the demand for rice keeps increasing. In this situation, damage to the rice plant, by any cause, is unacceptable. Rice plant disease detection is a very challenging issue [6]. Farmers can't detect disease, so they face great losses, and production could decrease due to disease. The country could

DOI: 10.1201/9781003145042-8

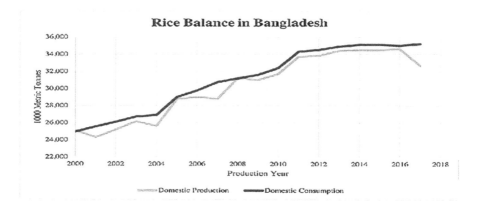

FIGURE 8.1 Rice production in Bangladesh [5].

also face great economic losses. The primary causes of plant disease are bacteria, fungi, viruses, and insects. There are various kinds of rice disease. Rice plant disease detection and its severity have always been challenging. Visual analysis is the only solution to detect rice plant disease; but most of time, we can't detect disease, because we cannot analyze the current situation of the plant and soil. Because visual investigation requires consistent human perceptions, the cycle (visual examination) is in general expensive, bulky, and tedious for enormous zones of plants. To cultivating larger harvests, we need first to last observation of the plant. It should start at the bottom of the problem. It should be created on plants or soil. If soils have a problem, then plants do, also [7]. Soilborne disease can create serious problems without showing any visual symptoms. Our device can detect dirt conditions through sensors and send the information to the web cloud. After that, our framework will calculate these data and show if there are any germs in the soil. The pH value is the most sensitive and significant factor for identifying infections precisely. A diverse kind of pH value speaks to the strength of the soil and the presence of microbes, organisms, or nematodes. At the point when the soil is influenced by these, then the pH, dampness, and temperature likewise are changed, and what is more, cause harm to seeds, roots, and soil nourishment. Figure 8.2 shows soilborne disease [8].

Paddy disease and its symptoms: More than 20 diseases are present in our country, of which 13 are major ones. Leaf smut, bacterial blight, and brown spot are most of the vital diseases. Those diseases are visible on the leaf by spot detection and changes in color.

Leaf smut: A typical symptom of leaf smut is the appearance of dark black or brown spots on the leaf. Small black linear lines may appear on leaf blades, and leaf tips may turn gray and dry.

Bacterial blight: This disease is well on its way to creating territories that have weeds and the stubble of contaminated plants. As a rule, the illness favors temperatures at 25°C–34°C, with relative humidity above 70%.

Rice Plant Disease Detection Using IoT

FIGURE 8.2 Soilborne disease.

Brown spot: The disease occurs in potassium deficient soils and areas with temperatures ranging from 25°C–28°C. It appears during the late growth stages of the rice crop, starting at the heading stage. Plants are most susceptible during panicle initiation onwards, and damage becomes more severe as plants approach maturity.

Three different diseases have characteristic patterns and shapes. The features of the diseases are described and illustrated in Figure 8.3 [9].

FIGURE 8.3 Different types of leaf disease [9].

122   Cybersecurity

To detect these types of disease, we need to monetize regularly. But during the cultivation phase, if we can follow those policies, then our losses will decrease. A device can help solve the problem, keeping an eye on the things that concern us. We use different types of sensors to create an IoT device and also make a framework to detect diseases. The device will show our current situation in the field. We can keep an eye out for all things at any place and any time. Our production cost will decrease, and profit will increase. We can address any kind of disease that occurs in the paddy field.

8.2   RELATED WORK

A researcher from Kerala, India, Amogh Jayaraj Rau, and his team chipped away at an IoT-based smart irrigation system and nutrient detection with disease analysis. In their study, temperature, mugginess, and photographs of plants were taken and shipped off to a Raspberry Pi. As the solenoid valve was straightforwardly associated with the Raspberry Pi, a farmer could turn the watering of the farm on and off. From the Raspberry Pi, that data was shipped off to the cloud, and the farmer could screen it on their cell phone [10].

S. V. Gaikwad and S. G. Galande present IoT implementation for remote checking of agrarian boundaries. A remote structure is made to screen regular conditions in the agriculture field, such as temperature, soil pH, soil wetness level, and leaf disease detection [11].

Numerous analysts have been chipping away at getting the natural information from ranches remotely. Yuta Kawakami and other scientists built a model for a rice cultivation support system equipped with a water-level sensor system. They outfitted their framework with a temperature and stickiness sensor, soil temperature sensor, soil water sensor, camera, and water-level sensor. Those data could be seen on an advanced cellphone or PC [12].

As per previous experience, human resource (HR) and disease specialists are expected to gather and dissect the information. In any case, numerous nations lack HR in farming, which is another huge issue. As of late, advancements such as drones have been actualized in agribusiness for various assignments, such as splashing pesticides and checking crops [13]. There are proposed sensor-based machine-to-machine (M2M) agribusiness-observing frameworks for agricultural nations, and other proposed IoT smart agribusiness devices with distributed computing, as in [14].

Plant diseases truly influence the typical development of plants, their yield, and the nature of agrarian items. From the overview introduced earlier, it is seen that the work on plant infection discovery utilizing the IoT is scant. Due to the abrupt change in climate of the atmosphere, the common climate of plant development has been harmed by pollution, and continuous cataclysmic events, at the same time as the improvement of agricultural innovation. In this chapter, the focus is on utilizing sensor-based innovation. We propose leaf disease detection utilizing the IoT, to help farmers innovatively. In the proposed work, the focus has been on the early discovery of disease and on disease on plant leaves [15].

Kaur and Kaur [16] aim to distinguish plant illnesses utilizing deep learning methods that will help farmers to rapidly and effectively distinguish illnesses, and thus would empower farmers to make legitimate strides right at the beginning. To accomplish greater precision in assessing a prescient model, the creators utilized a 10-overlay cross-approval procedure on their dataset. They utilized 2,589 unique pictures in performing tests and 30,880 pictures for preparing their model utilizing the Caffe profound learning structure. The precision expected of this model is 96.77%.

In this way, actualizing IoT design in farming, with drones, along with picture handling, is another test; it will give focal points to the agribusiness framework. The central point of utilizing drones rather than human work is that it can diminish time utilization for field study, carry out assignments in larger areas, equipped for handling all undertakings, and tell the client more quickly to prepare to shield their harvests from infection, which brings about greater profitability.

8.3 PROPOSED SYSTEM MODEL

Figure 8.4 is our proposed model. By accessing our previous research on paddy cultivation, we can find that temperature pH and moisture are very important for soil.

If this value undergoes a major change, that means our plant will be infected by soil disease or plant disease. So, f we assume those values are normal, then we can ensure that our plant is healthy. Otherwise, we should take the necessary steps. Similarly, when the plant is growing, we must follow up on the leaf. If leaf color is

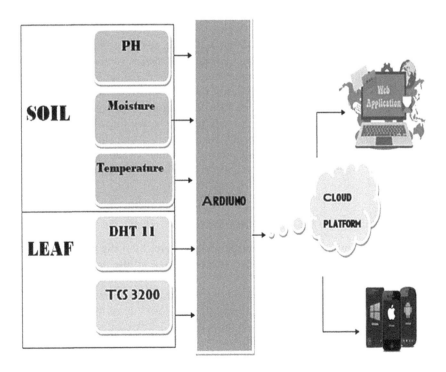

FIGURE 8.4 Arduino system model.

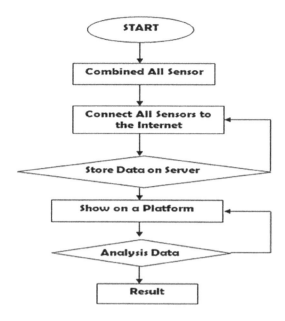

FIGURE 8.5 System flow chart.

changed or spotted and humidity becomes a chance, then we might think that the plant has some problem. To detect all these important things, we use various kinds of sensors to collect data. We use pH temperature and moisture sensors to collect data on soil, and also use TCS3200 and humidity for collecting data on leaves. TCS3200 detects leaf color, and the humidity sensor detects humidity. To collect all values, we use an Arduino Uno microcontroller by running C code. We find all sensor values in one platform. To send data to the Internet, we use an Ethernet shield. After sending data to the Internet we store it on an IoT platform, so that we can use this data for our future purposes. We can also access the data on any platform. Figure 8.5 provides our system flowchart.

8.3.1 FLOWCHART OF THE FOLLOWING SYSTEM

Figure 8.5 shows the flowchart of the following system. Start the following process one by one and complete all processes. After summing all the procedures, we will find our expected device.

After storing data on the ThingSpeak server, we can use that data for any purpose. We can analyze our data so that machines can detect that those measure values are good or bad in this situation. We compare it with the actual value that is good for plants, which we find through our study and research. After comparing those values, the device gives us a result of the current situation. Here is our algorithm to identify plant disease.

Hardware Section: The following hardware is used for this experiment.

FIGURE 8.6 Arduino UNO.

Arduino Uno: Arduino Uno is a microcontroller board based on the ATmega328P. It has 14 digital input and output pins, 6 analog input pins, 14 digital output pins, and a 16 MHz ceramic resonator (CSTCE16M0V53-R0), a USB connection, a power jack, an ICSP header, and a reset button, as shown in Figure 8.6. To collect sensor data, we use an Arduino Uno microcontroller. It's very easy to use. We configure it by using the C program.

Ethernet Shield: The Ethernet Shield is utilized to pass data from the Arduino board to the web through the wired web affiliation. Simply plug it into the Arduino board and interface with the Internet by the RJ45 link. All sensors are consolidated and related on the Arduino board, then it is related to the Ethernet Shield. The Arduino and Ethernet Shield have the same pin design, shown in Figure 8.7. It interfaces with Arduino on an SPI port. At the point when information goes through the Ethernet Shield, its LED light will blink consistently.

Moisture Sensor: The moisture sensor is used to detect the soil moisture content. This moisture sensor can be utilized to distinguish the dampness of soil or judge if there is water around the sensor, giving the plants access to your nursery connection for human assistance. When the dirt experiences a water shortage, the module yield is at a significant level; otherwise, the yield is at a low level. This sensor reminds the client to water their plants and screens the dampness of the soil, as shown in Figure 8.8. It has been generally utilized in agriculture, land-water systems, and plant cultivation.

DHT 11: The DHT11 is normally utilized for temperature and humidity sensors. DHT11 is an advanced temperature and moistness sensor to adjust computerized signal yield temperature and stickiness consolidated sensor. The sensor accompanies a devoted NTC to quantify temperature and an 8-digit microcontroller to yield the estimations of temperature and dampness as sequential information.

FIGURE 8.7 Ethernet shield.

FIGURE 8.8 Soil moisture sensor [19].

FIGURE 8.9 DH11 sensor.

Application-explicit modules catch innovation and computerized temperature and mugginess sensor innovation to guarantee high dependability and brilliant long-term strength, as shown in Figure 8.9.

pH sensor: pH represents the intensity of hydrogen, which is an estimation of the hydrogen particle focus in the body. The absolute pH scale goes from 1 to 14, with 7 viewed as neutral. A pH below 7 is considered acidic, and a pH more noteworthy than 7 is essential or soluble. The pH meter is a rational instrument that evaluates the hydrogen-molecule improvement in water-based game plans, showing its alkalinity or corrosiveness, presented as pH. The pH meter has two fundamental sections, one with a pointer that moves against a scale or an automated meter. With this sensor, you can measure the pH value (see Figure 8.10). We can easily connect this sensor with Arduino. If we want to measure pH values, then we need a pH sensor.

Soil Temperature Sensor: For measuring soil temperature we will be using the Dallas temperature in combination with the OneWire Arduino library. Those libraries communicate with one or multiple sensors. We can find the specifications and

FIGURE 8.10 pH meter [22].

FIGURE 8.11 DS18B20 temperature sensor [24].

FIGURE 8.12 TCS3200 color sensor [25].

information about the different types of DS18B20 sensors. Then, we will connect the sensor to the Arduino (see Figure 8.11).

TCS3200 Color Sensor: The TCS3200 Color Sensor Module has 4 LEDs, with a TCS3200 Color Sensor IC. The module is designed in such a way that 4 bright LEDs will light the object, and reflections from that object will strike the TCS3200 Color Sensor IC, to distinguish the color. This sensor is fundamentally used to recognize the shade of an article (see Figure 8.12). It has a variety of applications in industrial, medical, and consumer areas.

8.4 CIRCUIT DIAGRAM

Figure 8.13 demonstrates our trial setup and implanted framework, respectively. Here, we connect all sensors. We assemble circuits and obtain information by

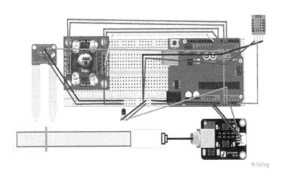

FIGURE 8.13 Experimental setup.

running the Arduino Code. Then, we connect the Ethernet Shield to send the data on a platform. We use the API key to connect the IoT platform and our Arduino board. After connecting, all data will be stored on the server. We can see our real-time data on this server dashboard. At first, we connected the Ethernet Shield with the Arduino board. Arduino and Ethernet Shields have the same pin configuration. We developed a board to connect from top to bottom with the same pin configuration. After connecting the Arduino and Ethernet Shield, we need to set up all sensors with Ethernet Shields one by one. We connect all sensors in analog and digital output pins. After connecting all sensors, we use Arduino IDE to generate Arduino code to find all sensor data. Then, we need to send it on to an IoT platform, to store the information for future data processing and analysis. We are implementing it in a rural area of Bangladesh and find outstanding results. It can be implemented anywhere in the world, at a very low cost. Here is our algorithm for identifying plant disease:

Input: The entire sensor values. Output: Plant is infected or not

Step 1: Collect all sensor value that we then store on the IoT platform

Step 2: Calculate all sensor values with a minimal value range. If those values are in the range then show (Soil condition is good. Leaf condition is good.) Otherwise, the soil may be affected.

Step 3: Finish (Nawaz et al. 2020).

8.5 RESULT

After running the Arduino code, we find all sensor values and represent them on a web and Android platform, so that we can visualize this data any time and any place. Figures 8.14 and 8.15 present a graph view of the results, which we find on the IoT platform.

The results and analysis can also be shown in an Android platform.

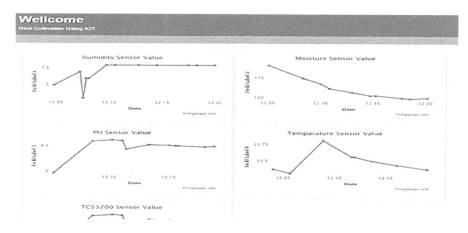

FIGURE 8.14 Data visualization on a website.

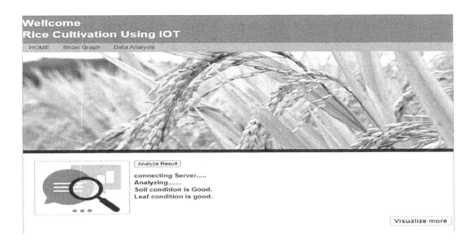

FIGURE 8.15 Data analysis.

8.6 CONCLUSION

In this paper, a framework is made to detect the nature of the leaves and soil. This work has a significant economic importance to Bangladesh. The proposed technique uses the sensor devices to perceive limits such as temperature, moisture, humidity, and the shade of the leaf, which are then differentiated with an instructive list to check whether the accumulated characteristics fall into a given range. This proposed model can provide extra benefits to farmers when they cultivate. For the same reason, they can increase production. If facing problems during cultivation, then they can take the necessary steps. The farmer can utilize their time and effort by using this device. This is an effective and cost-accommodating methodology for farmers, where they can gauge their plant climate utilizing this structure. We can quantify ongoing information by using those sensors and send that data by an Ethernet Shield through a web association and send that continuous information on an IoT platform. That way, we can identify it whenever and wherever and take the fundamental steps.

REFERENCES

[1] Bangladesh Rice Knowledge Bank. n.d. "Rice in Bangladesh." http://www.knowledge-bank-brri.org/riceinban.php (accessed Dec. 7, 2020).
[2] Daily Sun. 2020, May 15. "Bangladesh to become 3rd in global rice production-482142." https://www.daily-sun.com/post/482142/Bangladesh-to-become-3rd-in-global-rice-production (accessed Dec. 7, 2020).
[3] Trading Economics. n.d. "Bangladesh GDP from agriculture | 2006–2020 data | 2021–2022 forecast | Historical." https://tradingeconomics.com/bangladesh/gdp-from-agriculture (accessed Dec. 7, 2020).
[4] Ricepedia. n.d. "Bangladesh." http://ricepedia.org/bangladesh (accessed Dec. 7, 2020).
[5] A. M. Shew, A. Durand-Morat, B. Putman, L. L. Nalley, and A. Ghosh. 2019, May. "Rice intensification in Bangladesh improves economic and environmental welfare," *Environ. Sci. Policy*, vol. 95, pp. 46–57, doi: 10.1016/j.envsci.2019.02.004

[6] Ministry of Environment and Forests. 2012, May. Bangladesh: Rio + 20: National report on sustainable development. Government of Bangladesh. https://sustainabledevelopment.un.org/content/documents/981bangladesh.pdf. (accessed Dec. 07, 2020).

[7] M. C. Shurtleff, M. J. Pelczar, A. Kelman, and R. M. Pelczar. "Plant disease: | importance, types, transmission, & control." *Brittanica*. https://www.britannica.com/science/plant-disease (accessed Dec. 7, 2020).

[8] R. Ghosh, A. Tarafdar, D. Chobe, S. Chandran, U. Sudharani, and M. Sharma. 2019, June. "Diagnostic techniques of soil borne plant diseases: Recent advances and next generation evolutionary trends," *Biological Forum: An International Journal*, vol. 11, no. 2, pp. 1–13.

[9] S. Ramesh and D. Vydeki. 2020, June. "Recognition and classification of paddy leaf diseases using Optimized Deep Neural network with Jaya algorithm," *Information Processing in Agriculture*, vol. 7, no. 2, pp. 249–260, doi: 10.1016/j.inpa.2019.09.002.

[10] A. J. Rau, J. Sankar, A. R. Mohan, D. Das Krishna, and J. Mathew. 2017, July. "*IoT based smart irrigation system and nutrient detection with disease analysis*," in *2017 IEEE Region 10 Symposium (TENSYMP)*, pp. 1–4, doi: 10.1109/TENCONSpring.2017.8070100.

[11] S. V. Gaikwad and S. G. Galande. 2015. "Measurement of NPK, temperature, moisture, humidity using WSN." *International Journal of Engineering and Research*, vol. 5, no. 8, pp. 84–89.

[12] Y. Kawakami. 2016, January. "Rice cultivation support system equipped with water-level sensor system," *IFAC-Pap.*, vol. 49, no. 16, pp. 143–148, doi: 10.1016/j.ifacol.2016.10.027.

[13] Semantic Scholar. n.d. "Smart agriculture based on cloud computing and IOT." https://www.semanticscholar.org/paper/Smart-Agriculture-Based-on-Cloud-Computing-and-IOT-Tong-ke/fa7606fc2abe3d069efef6618d45da7a1dcb8bd5 (accessed Dec. 08, 2020).

[14] L. Karim, A. Anpalagan, N. Nasser, and J. Almhana. 2013. "Sensor-based M2M agriculture monitoring systems for developing countries: State and challenges." *Network Protocols and Algorithms*, vol. 5, no. 3, p. 68.

[15] R. Yakkundimath, G. Saunshi, and V. Kamatar. 2018. "Plant disease detection using IoT." *International Journal of Engineering Science and Computing*, vol. 8, no. 9, p. 5.

[16] R. Kaur and V. Kaur. 2018. "A deterministic approach for disease prediction in plants using deep learning." *International Journal of Computer Science and Mobile Computing*, vol. 7, no. 2, pp. 80–88.

[17] Arduino. n.d. "Arduino Uno Rev3 | Arduino official store." https://store.arduino.cc/arduino-uno-rev3 (accessed Dec. 8, 2020).

[18] Arduino. n.d. "Getting Started with the Arduino Ethernet Shield and Ethernet Shield 2." https://www.arduino.cc/en/Guide/ArduinoEthernetShield (accessed Dec. 8, 2020).

[19] SparkFun. n.d. "SparkFun soil moisture sensor - SEN-13322 - SparkFun electronics." https://www.sparkfun.com/products/13322 (accessed Dec. 8, 2020).

[20] Wikipedia. 2020, November 12. "Soil moisture sensor." https://en.wikipedia.org/w/index.php?title=Soil_moisture_sensor&oldid=988272240 (accessed Dec. 8, 2020).

[21] Dejan. 2016, January 13. "DHT11 & DHT22 Sensor Temperature and Humidity Tutorial," *HowToMechatronics*, https://howtomechatronics.com/tutorials/arduino/dht11-dht22-sensors-temperature-and-humidity-tutorial-using-arduino/ (accessed Dec. 8, 2020).

[22] RoboticsBD. n.d. "Analog pH sensor / Meter kit for arduino," https://store.roboticsbd.com/sensors/523-analog-ph-sensor-meter-kit-for-arduino-robotics-bangladesh.html (accessed Dec. 8, 2020).

[23] Khan Academy. n.d. "pH scale: Acids, bases, pH and buffers (article) |." https://www.khanacademy.org/science/biology/water-acids-and-bases/acids-bases-and-ph/a/acids-bases-ph-and-bufffers (accessed Dec. 8, 2020).

[24] Components101. n.d. "*DS18B20 temperature sensor*". https://components101.com/sensors/ds18b20-temperature-sensor (accessed Dec. 8, 2020).

[25] ProjectShopBD. n.d. "GY-31 TCS230 TCS3200 color sensor module / color recognition sensor module #403 |." http://projectshopbd.com/product/gy-31-tcs230-tcs3200-color-sensor-module-color-recognition-sensor-module-403/ (accessed Dec. 8, 2020).

9 Secure Protocols for Biomedical Smart Devices

Poonam Sharma
University Institute of Computing, India

Prabhjot Kaur
GGSIP University, India

Kamaljit Singh Saini
University Institute of Computing, Chandigarh University, India

CONTENTS

9.1 Introduction 132
9.2 Communication Architecture in Smart Devices 133
9.3 Overview of Biomedical Smart Devices 133
    9.3.1 Health-Oriented Smart Watch 134
    9.3.2 Blood Pressure Monitor 134
    9.3.3 Wireless Smart Glucometer 134
    9.3.4 Brain-Sensing Headband 134
    9.3.5 Smart Temporal Thermometer 134
    9.3.6 Wearable ECG Monitors 135
    9.3.7 Heart Rate Sensors 135
    9.3.8 Pulse Oximeter Sensors 135
    9.3.9 Motion Sensors 135
9.4 Security Requirements for Communication in Biomedical Smart Devices ... 135
    9.4.1 Data Confidentiality 135
    9.4.2 Scalability 135
    9.4.3 Data Integrity 136
    9.4.4 Data Authenticity 136
    9.4.5 Data Availability 136
    9.4.6 Data Security 136
    9.4.7 Data Confidentiality 136
    9.4.8 Data Privacy 137
    9.4.9 Data Freshness 137

DOI: 10.1201/9781003145042-9

9.4.10	Secure Management	137
9.4.11	Dependability	137
9.4.12	Secure Localization	137
9.4.13	Accountability	137
9.4.14	Flexibility	137

9.5 Threats and Attacks 137

9.5.1	Replayed/Spoofed Routing Information	138
9.5.2	Selective Forwarding	138
9.5.3	Sinkhole Attacks	138
9.5.4	Sybil Attacks	138
9.5.5	Wormholes	138
9.5.6	HELLO Flood Attacks	138
9.5.7	Replay Attack	138
9.5.8	Denial of Service Attack	139
9.5.9	Man-in-the-Middle Attack	139
9.5.10	Flooding	139
9.5.11	Jamming	139
9.5.12	Tampering	139

9.6 Application Area for Smart Devices in Medical Health Care System 139
9.7 Security Protocols for Smart Devices 140

9.7.1	Robust and Efficient Energy Harvested Aware Routing Protocol	140
9.7.2	Lightweight Information Encryption Protocol	140
9.7.3	A Secure Protocol for User Authentication and Key Agreement	140
9.7.4	Node-to-Node Authentication Protocol by Eliminating the Man-in-Middle Attack	141
9.7.5	Lightweight Anonymous Authentication Protocol	141
9.7.6	Lightweight Data Confidentiality and Authentication Protocols (2012)	141
9.7.7	A Trust Key Management Protocol	141
9.7.8	Physiological-Signal-Based Key Agreement Protocol	142
9.7.9	Localized Encryption and Authentication Protocol (LEAP)	142
9.7.10	Random Key Predistribution Schemes	142

9.8 Other Security Mechanisms for Smart Devices 142
9.9 Conclusion 145
References 145

9.1 INTRODUCTION

Recent advances in information technology have provided a path to wireless sensor networks. This type of network is a collection of a copious amount of smart devices, also known as sensor nodes. Each device has its own power unit, sensing unit, processing unit, and communication unit. Biomedical smart devices, in collaboration with cloud computing and the Internet of Things (IoT), form a new technology for e-health care systems [1]. These devices can collaborate with one another using wireless communication techniques, and they can remotely monitor the real-time parameters of patients' physiology and movements, as collected by small wearable sensor devices [2]. The sensor nodes [3] can be classified as:

Implant Node: These nodes are implanted under a person's skin.

On Body Surface Node: These types of nodes are situated on the body of a person.

External Node: These types of nodes are placed at least 5 m away from the human body.

9.2 COMMUNICATION ARCHITECTURE IN SMART DEVICES

Communication architecture [4] includes three layers, as shown in Figure 9.1. The application layer collects health-care-related data via sensor devices. The network layer transmits the health-related data obtained from the application layer to the next layer. Finally, the physical layer stores health-related data for medical services purposes.

9.3 OVERVIEW OF BIOMEDICAL SMART DEVICES

In the present era of technology, there are advancements in medical health care systems, also. Technology provides us a copious amount of health-related smart devices, which are used to monitor various parameters of the human body, as shown in Figure 9.2. Some of them are [42]:

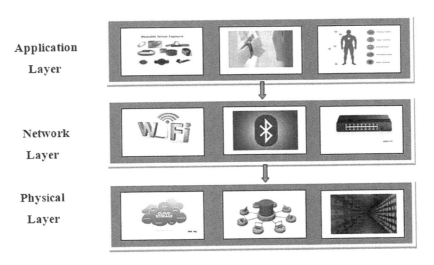

FIGURE 9.1 Communication architecture of smart devices.

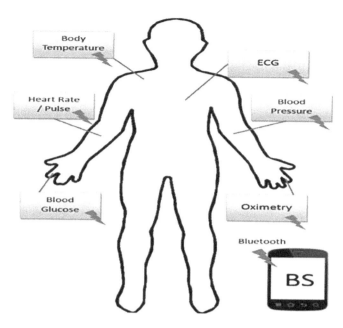

FIGURE 9.2 Biomedical smart devices [7].

9.3.1 Health-Oriented Smart Watch

This device keeps track of heartbeat irregularities during sleep and measures the blood oxygen level.

9.3.2 Blood Pressure Monitor

These devices are used to monitor the blood pressure level in the human body.

9.3.3 Wireless Smart Glucometer

Today, many people suffer from diabetes. These smart devices measure glucose levels in the blood of people with diabetes and can be operated every day, very easily.

9.3.4 Brain-Sensing Headband

This device monitors brain activity. These types of sensors are widely used for physically disabled patients, to understand their brain activity.

9.3.5 Smart Temporal Thermometer

These devices are used to measure human body temperatures, to evaluate physical states. Body temperature might be dependent upon physical activity, but abnormal

Secure Protocols for Biomedical Smart Devices 135

temperatures may be caused by disease. And the temperature measure must be accurate.

9.3.6 WEARABLE ECG MONITORS

These devices are used to measure an electrocardiogram. Pressure sensors are used to monitor ECG, which are connected above the adult carotid artery, to record the pulse wave.

9.3.7 HEART RATE SENSORS

This type of sensor is used to measures the pulse waves of the human body. It also provides convenient information on possible cardiovascular diseases, which are detrimental to human beings.

9.3.8 PULSE OXIMETER SENSORS

These sensors are used to measure the oxygen level of blood.

9.3.9 MOTION SENSORS

These sensors are used to track physical activity and the movement of the patient's body, to ensure good health and correct posture.

9.4 SECURITY REQUIREMENTS FOR COMMUNICATION IN BIOMEDICAL SMART DEVICES

Security is the paramount issue, and it should not be ignored. In an e-health care system, during the communication between devices, the information related to a patient's medical diagnosis and treatment is sensitive. If the information is hacked during the communication between a base station and a physician, this could directly affect patients' medical treatment. Sometimes, their life depends upon this e-health care system. So, to assure secure communication in smart devices there are security and privacy requirements [5], as mentioned in Figure 9.3, which should be followed during the transmission of information.

9.4.1 DATA CONFIDENTIALITY

For the confidentiality of a patient's sensitive information, the data must be protected from illegal and unauthorized access. Several encryption and decryption methods have been proposed, for data confidentiality.

9.4.2 SCALABILITY

In an e-health care system, patient-related data could be shared between many physicians for health-related consultations. This energy consumption must be considered.

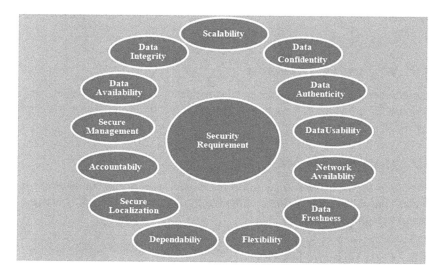

FIGURE 9.3 Security requirements in smart devices.

9.4.3 Data Integrity

Data integrity means protecting the accuracy of data. It is a process to ensure that receivers received exactly the same data delivered by the sender. The data received should not be modified by unauthorized access during the transmission.

9.4.4 Data Authenticity

Authentication is a major requirement for medical health systems. Every sensor node in the network must have its unique identity during the communication.

9.4.5 Data Availability

Data availability means that a patient's lifesaving information must be available in a health care system network every time, so that physicians can access it from anywhere. As a consequence, patients can be treated immediately in the event of an emergency.

9.4.6 Data Security

The physical layer, which stores health-related information, must be secure from unauthorized access. To ensure data security, various encryption techniques are used.

9.4.7 Data Confidentiality

Confidentiality means the privacy of patients' health-related data. This data must be protected from unauthorized access that could, in some cases, even be detrimental to their life.

Secure Protocols for Biomedical Smart Devices

9.4.8 DATA PRIVACY

Patients' health-related data can only be accessed by authorized persons. This is sometimes known as authorization.

9.4.9 DATA FRESHNESS

During the transmission, data freshness must be considered as a primary need, to maintain the integrity and confidentiality of data. Data freshness means that data can neither be a recorded copy by an adversary nor be older, which could create confusion in the network.

9.4.10 SECURE MANAGEMENT

Several key distribution schemes can be consider for secure management during the communication in a sensor network.

9.4.11 DEPENDABILITY

This means the data is reliable. In other words, data must be correct to avoid any error during the transmission. To provide error-free communication, various error-correcting code techniques have been proposed.

9.4.12 SECURE LOCALIZATION

Localization means the correct location of patients. A patient's location must be secure from illegal tracking by hackers.

9.4.13 ACCOUNTABILITY

The medical health care service provider must keep track of any misuse of any medical equipment's data and all health-related data.

9.4.14 FLEXIBILITY

In term of access control, patient data must be flexible for the hospital, as well as for a physician. In case of emergency, data must be accessible by different physicians in different hospitals.

9.5 THREATS AND ATTACKS

Smart devices are vulnerable to an enormous number of attacks and threats. These attacks sometime lead to death of patients, due to false information transmitted as a patient's data by an adversary. Figure 9.4 is a pictorial representation of such a situation. In this section, we discuss some attacks on different layers [47] of communication: physical layer attacks, network layer attacks, and transport layer attacks.

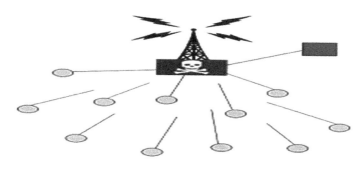

FIGURE 9.4 HELLO flood attack [18].

9.5.1 Replayed/Spoofed Routing Information

An attacker can create fake error messages and increase end-to-end delay.

9.5.2 Selective Forwarding

In this attack, some of the information is not forwarded by the nodes. This results in the loss of information, because malicious nodes can delay or stop the transmission of information by refusing to forward it [7].

9.5.3 Sinkhole Attacks

In this attack, all the traffic of a specific area is forced to pass through an arbitrary node.

9.5.4 Sybil Attacks

In this attack, multiple identities are presented by a single node to other nodes.

9.5.5 Wormholes

In this attack, messages can be captured by attacker and then replayed to other nodes.

9.5.6 HELLO Flood Attacks

An attacker induced a hallucination in which every node of a network thought they all were neighbors.

9.5.7 Replay Attack

A forwarded packet is copied by an attacker who intercepts the packet and then can send modified information to the receiver, with the purpose of delaying and misleading the receiver with the updated information. This attack lowers network performance and blocks transmission.

Secure Protocols for Biomedical Smart Devices

9.5.8 Denial of Service Attack

This is the most common attack in a wireless sensor network. In this case, an attacker consumes all the resources and forwards various false requests to make the network unavailable (the network is unable to provide the intended services) for certain users.

9.5.9 Man-in-the-Middle Attack

A connection is established within the nodes in the network and starts communication with a delusion as a part of the same network.

9.5.10 Flooding

An attacker establishes a new connection and afterward it requests continuously.

9.5.11 Jamming

This is a kind of Denial of Service attack, which denies service to a node by sending a jamming signal with the 2.4 GHz frequency, which makes the wireless network nonfunctional [8].

9.5.12 Tampering

This is a node capture attack. This attack takes complete control over the captured node and damages or modifies node service. The cryptography technique is highly susceptible to tampering attacks.

9.6 APPLICATION AREA FOR SMART DEVICES IN MEDICAL HEALTH CARE SYSTEM

There are number of innovative and interesting applications for smart devices that are widely used in medical systems to provide e-health facilities to patients. These applications include:

- Telemedicine is one of most demanding application of smart devices, which provide remote facility to medical professionals to control the health situations of their patients [9].
- Rehabilitation can detect human movement by real-time feedback for patients [10].
- Assisted living/biofeedback measures the physiological parameters of the human body by providing medical feedback to the respective user.
- Remote health care monitoring systems help to measure the heartbeat rate, blood pressure, and accelerometer data remotely, to monitor patient's health-related data.

9.7 SECURITY PROTOCOLS FOR SMART DEVICES

Both security protocols and algorithms are security mechanisms designed to prevent attacks. Sensitive information in a medical health care system must be protected from unauthorized access. For this purpose, security protocols for biomedical smart devices have been proposed by the researcher to provide security in health care systems.

9.7.1 ROBUST AND EFFICIENT ENERGY HARVESTED AWARE ROUTING PROTOCOL [11]

In 2019, a Robust and Efficient Energy Harvested Aware Routing Protocol was proposed, with a clustering algorithm to analyze energy consumption in different body modes. In this protocol, the best forwarder nodes are chosen, based on cost function (CF). The data communication and authentication process between the smart devices and data center are divided into five phases: the Deployment Phase, Initialization Phase, Cost Function Calculation Phase, Scheduling Phase, and Data Transmission Phase. In the first phase, a total of 14 sensors are deployed on the human body, to sense the physiological parameters. In the second phase, two-way transmission begins by sending a hello message. In the third phase, loads on sink nodes are calculated based on cost function and forward node selection is performed based on the best cost function. The scheduling phase schedules the process of transmission through multi-hope nodes. In the last phase, data transmission starts by adopting multi-hope communication, which consumes less energy. It is designed with the intent to provide secure communication in smart devices.

9.7.2 LIGHTWEIGHT INFORMATION ENCRYPTION PROTOCOL [12]

In 2018, a lightweight information encryption mechanism was proposed for medical environments to increase the data security at minimum cost, based on channel characteristics to extract real-time keys. For the key distribution, key extraction, inconsistency removal, and the fuzzy key unification method are adopted. This protocol is similar to RSS to authenticate nodes. In this protocol, communication between a coordinator node and a sensor node takes place into two steps: The first phase is the node authentication phase, to authenticate the node. The second is the encryption information phase, for which n-dimension quantifier and fuzzy extraction methods are proposed. In this method, key extraction techniques are used, instead of key distribution, to reduce the cost. The method proposed provides security against tampering attacks. This attack takes complete control over the capture node and damages or modifies node services.

9.7.3 A SECURE PROTOCOL FOR USER AUTHENTICATION AND KEY AGREEMENT [13]

In 2017, a mechanism for key distribution based on mutual authentication process was proposed. The mechanism is divided into three parts: a user registration process, an authentication process, and a key agreement process. In this mechanism, a registration center (RC) plays a role in key distribution, like that of a base station (BS).

Secure Protocols for Biomedical Smart Devices

The first process begins with the user registration, followed by a log-in step. In the second process, authentication is performed between a gateway and sensors to provide secure communication between users and smart devices. In the last process, key distribution is performed by using the hashing concept. This protocol provides security against Denial of Service attacks, which make the network unavailable for legitimate users, and from Spoofing Attacks.

9.7.4 NODE-TO-NODE AUTHENTICATION PROTOCOL BY ELIMINATING THE MAN-IN-MIDDLE ATTACK [14]

This protocol was designed to provide secure communication in smart device environments. This protocol was proposed in 2015, using a data encryption technique, the Data Encryption Algorithm (DEA), and divides the entire process into four phases by using 52 keys for encryption. Communication occurs in four phases, starting with an initialization phase. After that, a key generation phase starts, in which a secret key is generated by the BS (base station). The cluster head receives this secret key and forwards it to the sensor node. The third phase is communication, which begins if keys are same for both sides; otherwise, the communication is refused. At the end is a session expired phase, in which a session is expired by the BS. After that, a new session begins.

9.7.5 LIGHTWEIGHT ANONYMOUS AUTHENTICATION PROTOCOL [15]

In 2015, a secure IoT-based health care system was proposed with the contribution of a body sensor network (BSN) to accomplish security requirements. This protocol was proposed for health care systems to monitor patient health conditions. This protocol consists of two phases, beginning with a registration process. In the second phase, a lightweight anonymous authentication protocol for security is proposed, using encryption techniques.

9.7.6 LIGHTWEIGHT DATA CONFIDENTIALITY AND AUTHENTICATION PROTOCOLS (2012) [16]

A three-layered network architecture for the Wireless Body Sensor Network (WBSN) was proposed in 2012. To limit energy consumption and key distribution complexity, in the proposed scheme, sensors and the BS utilize three keys for secure communication. In a Wireless Body Area Network (WBAN) environment in 2012, a lightweight data confidentiality and authentication protocol was proposed to limit the energy consumption and key distribution complexity. In this protocol, sensors and the BS utilized three keys for secure communication in a smart device environment.

9.7.7 A TRUST KEY MANAGEMENT PROTOCOL [17]

In 2011 a trust key-based management protocol for smart devices was proposed to provide security and privacy for sensor data. The communication process between sensor nodes and the base station perform in four phases.

142   Cybersecurity

9.7.8   Physiological-Signal-Based Key Agreement Protocol [18]

A secure key distribution scheme PSKA (physiological-signal-based key agreement) was proposed in 2010. In this scheme, two sensors agree upon a single key. The analysis of the proposed scheme shows that a patient's physiological signals can meet the security requirement.

9.7.9   Localized Encryption and Authentication Protocol (LEAP) [19]

This protocol proposed a secure key distribution mechanism to provide confidentiality and authentication in smart device networks. A total of four different types of keys are used for information exchange. First is an individual key, which is used for secure communication between node and the BS. Second is a pairwise key, which is used to secure locally broadcast messages shared with another sensor node. Third is a cluster key [20,21], which is used to provide privacy and source authentication. Last is a group key, which is globally shared by the BS for encrypting messages that are broadcast to the whole group.

9.7.10   Random Key Predistribution Schemes [22]

This protocol has four phases. First is the key predistribution phase: A centralized key server generates which provide a large key pool offline. The procedure for offline key distribution is as follows:

- Assign a unique node identifier or key ring identifier to each sensor.
- Select "m" different keys for each sensor from the key pool to form a key ring.
- Load the key ring into the memory of the sensor.

Second is the sensor deployment phase: The sensors are randomly chosen and uniformly distributed in a large area. Typically, the number of neighbors of a sensor (n) is much smaller than the total number of deployed sensors (N). Third is the key discovery phase: Each sensor broadcasts its key identifiers in clear-text or uses a private share-key discovery scheme to discover the keys shared with its neighbors. By comparing the possessed keys, a sensor can build a list of reachable nodes with which to share the keys and then broadcast its list. Using the lists received from neighbors, a sensor can build a key graph based on the key-share relations among neighbors. Fourth is the pairwise key establishment phase: If a sensor shares key(s) with a given neighbor, the shared key(s) can be used as their pairwise key(s). If a sensor does not share key(s) with a given neighbor, the sensor uses the key graph built during the key discovery phase to find a key path, to set up the pairwise key [23–25].

9.8   OTHER SECURITY MECHANISMS FOR SMART DEVICES

Besides the aforementioned protocols, there are other key mechanisms [43–46] for smart devices that can also be considered as having security purposes:

- **Privacy and device authentication scheme:** a secure anonymous device authentication scheme is proposed for biomedical purposes. In the proposed scheme, data communication is performed in three phases: an initialization phase, a registration phase, and an authentication and key establishment phase, by using the JPBC library [26].

- **A secure architecture design for sensor devices (smart devices):** WBSN architecture based on an RSA algorithm and a storage site, from which data can be accessed safely. This research talks about how to impart connections between the patient and the physician and build up a few plans to accomplish a safe and ensured correspondence between them. In this proposed system, the WBSN system will work in two stages. Any correspondence between the patient and the physician isn't immediate, however it will happen in two phases: the primary stage between the patient and the storage site, and the second phase between the storage site and the physician [27].
- **A secure communication framework:** Here, the author proposes a hybrid algorithm for heterogeneous sensor networks. The propose work provides secure communication between a sink node and BS. The author divides the communication into five phases and uses an elliptic curve Diffie–Hellman algorithm for key exchange [28].
- **Secure and low energy consumption scheme**: The author uses the Blowfish algorithm and proposes a security and low energy consumption scheme in WBANs. The author divides the communication process in two phases: key expansion (uses a recursive key) and data encryption (uses symmetric encryption) [29].
- **Design homomorphism encryption structure framework:** For secure communication, this health care system structure gathers therapeutic information from WBANs, transmits them through a broad remote sensor network, and finally distributes them into remote individual zone systems (WPANs) by means of a portal. HES includes two schemes: (a) a GSRM (groups of send-receive model) plan to acknowledge key dissemination and secure information transmission, and (b) an HEBM (homomorphism encryption based on matrix) plan to guarantee protection and a specialist framework ready to break down the mixed restorative information and criticize the outcomes naturally. Experimental evaluations are performed to show the security, protection, and enhanced execution of HES, compared to current frameworks or schemes. Finally, the model execution of HES is investigated to check its feasibility. [30]
- **A lightweight Authentication framework**: Instead of a digital signature, the author uses the concept of hash chains to propose an access control framework. It follows four phases for key distribution in a secure way and a new node injection phase to avoid man-in-the-middle attacks. [31]
- **A lightweight algorithm based on CHAOS scrambling for security:** a CHAOS-based algorithm is proposed to provide security in WBANs in case of emergency. The author uses Chaos-based cryptography and ECG data for analysis purposes [32].
- **Design a lightweight encryption algorithm (LEA):** This algorithm provides secure communication between the application layer and the transmission layer for a patient's data transmission in health care systems. The

communication process is divided into two parts: key scheduling and encryption and decryption [33].

- **KNAPSACK Encryption Algorithm**: An encryption algorithm is proposed to provide high accuracy for a patient's data. Cryptography is used with the concept of compression and decompression. In the proposed algorithm, data is compressed before encryption and decompressed during the entire communication from sensor nodes to data storage [34].
- **Design an intrusion detection system:** A distributed intrusion detection system is proposed, based on a genetic algorithm that provides security from jamming attacks with low energy consumption [35].
- **Design an OTP-based framework:** For secure information transmission, the author introduces an idea to use OTPs in WSN for secure data transfer, by reinstalling some OTPs on sensors devices. The author designed an OTP-based framework for secure information transmission in WSN [36].
- **A productive certificate less public auditing (CLPA) scheme**: This is a cryptographic technique that can remotely verify the integrity of information stored. The main objective of a CLPA scheme is to protect the user's master key [37].
- **Secure key management scheme by utilizing SIM card number:** A secure key management scheme is proposed based on an ECC (elliptic curve cryptography) algorithm. The proposed scheme follows three phases for secure key distribution and also utilizes a SIM card number as a unique identification code situated on smart devices [38].
- **A secure key based on spatio-temporal characteristics of remote channel**: A system is proposed to protect information by using the spatiotemporal characteristics of the remote channel used for correspondence. The proposed arrangement empowers two parties to produce firmly coordinating connection fingerprints, which connect an information session with a remote connection. These fingerprints are lightweight and empower some security properties, for example, responsibility, no revocation, and opposing man-in-the-middle attacks. The proposed system is tested utilizing body-worn sensors [39].
- **Key management technique based on threshold cryptography:** The author proposes a key management technique in WSN. In this technique, the author uses threshold cryptography. Encryption and decryption procedures are utilized for data transmission in the system. On the behalf of all sensing nodes, the data communication is performed by forwarding nodes to the BS. Here, the BS plays the role of key distributor [40].
- **Biometric-based secure framework for medical care system:** The author utilizes the concept of biometrics to design a security framework for WBANs. In the proposed scheme, the author uses biometric data for authentication purposes and also in an encryption mechanism as a shared key. Experimental outcomes show that the proposed methodology could accomplish precise verification execution without the additional prerequisite of a key [41].

9.9 CONCLUSION

Biomedical Devices have become more popular in medical health care systems to monitor patient health-related parameters. Although these smart devices are designed to provide medical help to older people or to provide help to people who are suffering from some disease, they can become vulnerable in terms of security. So, security mechanisms should be implemented while using them, to avoid precarious situations in future. In this chapter, we present an overview of biomedical devices and their security issues, with some current solutions.

REFERENCES

[1] R. Sh, "Wireless Body Area Networks: An Overview," *International Research Journal of Engineering and Technology (IRJET)*, Vol. 4, Issue 5, pp. 1356–1360, May 2017.

[2] J. J. P. C. Rodrigues, D. B. R. Segundo, H. A. Junqueira, M. H. Sabino, R. M. Prince, J. Muhtadi, and V. C. De Albuquerque, "Enabling Technologies for the Internet of Health Things," *IEEE Access*, Vol.6, pp. 13129–13141, Jan. 2018.

[3] S. Movassaghi, M. Abolhasan, J. Lipman, D. Smith, and A. Jamalipour, "Wireless Body Area Networks: A Survey," *IEEE Communications Surveys & Tutorials*, Vol. 16, Issue 3, pp. 1658–1686, Jan. 2014.

[4] Y. Ma, Y. Wang, J. Yang, Y. Miao, andW. Li, "Big Health Application System based on Health Internet of Things and Big Data," *IEEE Access*, Vol. 5, Issue 99, pp. 7885–7897, June 2017.

[5] M. Sheraz, A. Malik, M. Ahmed, T. Abdullah, N. Kousar, M. Nigar Shumaila, and M. Awais, "Wireless Body Area Network Security and Privacy Issue in E-Health care," *(IJACSA) International Journal of Advanced Computer Science and Applications*, Vol. 9, Issue 4, pp. 209–215, 2018.

[6] S. Janabi, I. Shourbaji, M. Shojafar, and S. Shamshirband, "Survey of Main Challenges (Security and Privacy) in Wireless Body Area Networks for Healthcare Applications," *Egyptian Informatics Journal*, Vol.18, Issue 2, pp. 113–122, 2016.

[7] M. Simran, V. Kaur, and P. Singh, "A Secure Authentication Mechanism for Wireless Sensor Networks," *IOSR Journal of Engineering (IOSRJEN)*, Vol. 08, Issue 9, pp. 61–67, Sept. 2018.

[8] P. Niksaz, "Wireless Body Area Networks: Attacks and Countermeasures," *International Journal of Scientific & Engineering Research*, Vol. 6, Issue 9, pp. 556–568, Sept. 2015.

[9] R. Khan, and A. K. Pathan, "The State-of-the-Art Wireless Body Area Sensor Networks: A Survey," *International Journal of Distributed Sensor Networks*, Vol. 14, Issue 4, March 2018.

[10] R. Negraa, I. Jemilia, and A. Belghith, "*Wireless Body Area Networks: Applications and Technologies*," in *The Second International Workshop on Recent Advances on Machine to Machine Communications*. Procedia Computer Science, pp. 1274–1281, 2016. Madrid, Spain: Elsevier.

[11] Z. Ullah, I. Ahmed, T. Ali, N. Ahmad, F. Niaz, and And Y. Cao, "Robust And Efficient Energy Harvested-aware Routing Protocol with Clustering Approach in Body Area Networks," *IEEE Access*, Vol. 7, pp. 33906–33920, March 2019.

[12] P. Zhang, and J. Ma, "Channel Characteristic Aware Privacy Protection Mechanism in WBAN,", *Article Sensors 2018.*Vol 8, Issue 8, pp. 1–15, 2018.

[13] G. Choi, and I. Lee, *"A Key Distribution System for User Authentication using Pairing-based in a WSN,"* in *4th International Conference on Computer Applications and Information Processing Technology (CAIPT)*, pp. 1–4, Aug. 2017. Kuta Bali, Indonesia: IEEE.

[14] P. Joshi, M. Verma, and P. R. Verma, "Secure Authentication Approach using Diffie-Hellman Key Exchange Algorithm for WSN," in *International Conference on Control, Instrumentation, Communication and Computational Technologies (ICCICCT)*, pp. 527–531, Dec. 2015. Kanyakumari, India: IEEE.

[15] P. Gope, and T. Hwang, "BSN-Care: A Secure IoT-based Modern Healthcare System Using Body Sensor Network," *IEEE Sensors Journal*, Vol. 16, Issue 5, pp. 1368–1376, Nov. 2015.

[16] P. K. Sahoo, "Efficient Security Mechanisms for m-Health Applications using Wireless Body Sensor Networks," *Sensors*, Vol.12, Issue 9, pp. 12606–12633, Feb. 2012.

[17] M. Mohammed, F. Mohammed, and B. Boucif, "Trust Key Management Scheme for Wireless Body Area Networks," *International Journal of Network Security*, Vol. 12, Issue 2, pp. 75–83, Mar. 2011.

[18] K. K. Venkatasubramanian, A. Banerjee, and S. K. S. Gupta, "PSKA: Usable and Secure Key Agreement Scheme for Body Area Networks," *IEEE Transactions on Information Technology in Biomedicine*, Vol. 14, Issue 1, pp. 60–68, Jan. 2010.

[19] C. Intanagonwiwat, R. Govindan, and D. Estrin, *"Directed Diffusion: A Scalable and Robust Communication Paradigm for Sensor Networks,"* in *Proceedings of MobiCOM'00*, pp. 56–67, August 2000. Boston, Massachussetts: ACM.

[20] C. Karlof, Y. Li, and J. Polastre, ARRIVE: "An Architecture for Robust Routing in Volatile Environments." Technical Report UCB/CSD-03-1233, University of California at Berkeley, March 2003.

[21] S. Madden, R. Szewczyk, M. Franklin, and D. Culler. *"Supporting Aggregate Queries over Ad-Hoc Wireless Sensor Networks,"* in *4th IEEE Workshop on Mobile Computing Systems & Applications*, pp. 49–58, June 2002. Callicoon, NY, USA: IEEE.

[22] L. Eschenauer, and V. D. Gligor, *"A Key-Management Scheme for Distributed Sensor Networks,"* in *Proceedings of 9th ACM Conference on Computer and Communication Security (CCS-02)*, pp. 41–47, Nov. 2002. New York, USA: ACM.

[23] H. Chan, A. Perrig, and D. Song, *"Random Key Predistribution Schemes for Sensor Networks,"* in *Proceedings of 2003 Symposium on Security and Privacy*, pp. 197–215, May 11–14, 2003. Los Alamitos, CA: IEEE Computer Society.

[24] D. Liu, and P. Ning, *"Establishing Pairwise Keys in Distributed Sensor Networks,"* in *Proceedings of 10th ACM Conference on Computer and Communications Security (CCS'03)*, Oct. 2003, pp. 52–61. Washington, DC, USA: ACM.

[25] W. Du, J. Deng, Y. S. Han, and P. K. Varshney, *"A Pairwise Key Pre-distribution Scheme for Wireless Sensor Networks,"* in *Proceedings of 10th ACM Conference on Computer and Communications Security (CCS'03)*, pp. 42–51, Oct. 2003. Washington, DC, USA: ACM.

[26] V. Odelu, S. Saha, R. Prasath, L. Sadineni, M. Conti, and M. Jo, " Efficient Privacy Preserving Device Authentication in WBANs for Industrial e-Health Applications," *Computers & Security*, Vol. 83, pp. 300–312, Mar. 2019.

[27] K. Sweekat, F. Zarka, B. Maala, and S. Ahmad, "A New Secure Wireless Body Sensor Network Architecture," *International Journal of Engineering Trends and Applications (IJETA)* Vol. 5, Issue 1, pp. 41–45, Jan. 2018.

[28] S. Farooq, D. Prashar, and K. Jyoti, "Hybrid Encryption Algorithm in Wireless Body Area Networks (WBAN)," in *Intelligent Communication, Control and Devices, Advances in Intelligent Systems and Computing 624*, Dehradun, India: Springer Nature, pp. 401–410, 2018.

Secure Protocols for Biomedical Smart Devices **147**

[29] C. R. Rani, L. S. Jagan, C. L. Harika, V. V. Durga, and R. Amara, "Light Weight Encryption Algorithm for Wireless Body Area Networks," *International Journal of Engineering Technology*, Vol. 7, Issue 2, pp. 64–66, 2018.

[30] H. Huang, T. Gong, N. Ye, R. Wang, and Y. Dou, "Private and Secured Medical Data Transmission and Analysis for Wireless Sensing Healthcare System," *IEEE Transactions on Industrial Informatics*, Vol. 13, Issue 3, pp. 1227–1237, June 2017.

[31] H. Moon, U. Iqbal, and G. M. Bhat, *"Light Weight Authentication framework for WSN,"* in *International Conference on Electrical, Electronics, and Optimization Techniques (ICEEOT)*, pp. 3099–3105, Mar. 2016. Chennai, India: IEEE.

[32] S. F. Raza, *"A Proficient Chaos Based Security Algorithm for Emergency Response in WBAN System,"* *IEEE Students' Technology Symposium*, pp. 18–23, Sep. 2016. Kharagpur, India: IEEE.

[33] A. Z. Alshamsi, E. S. Barka, and M. A. Serhani, *"Light Weight Encryption Algorithm in Wireless Body Area Networks for e-Health Monitoring,"* in *12th International Conference on Innovations in Information Technology (IIT)*, pp. 144–150, Nov. 2016. Al Ain, Abu Dhabi: IEEE.

[34] N. A. Anoop, and T. K. Parani, *"A Real Time Efficient & Secure Patient Monitoring Based on Wireless Body Area Networks,"* in *Online International Conference on Green Engineering and Technologies (IC-GET)*, pp. 1–5, Nov. 2016. Coimbatore, India: IEEE.

[35] G. Thamilarasu, *"Genetic Algorithm Based Intrusion Detection System for Wireless Body Area Networks,"* *IEEE Symposium on Computers and Communication (ISCC)*, pp. 160–165, July 2015. Larnaca, Cyprus: IEEE.

[36] F. Busching, and L. Wolf, "The Rebirth of One-Time Pads—Secure Data Transmission from BAN to Sink," *IEEE Internet of Things Journal*, Vol. 2, Issue 1, pp. 63–71, Feb. 2015.

[37] D. He, S. Zeadally, and L. Wu, "Certificateless Public Auditing Scheme for Cloud-Assisted Wireless Body Area Networks," *IEEE Systems Journal*, Vol. 12, Issue 1, pp. 64–73, May 2015.

[38] Y. s. Lee, E. Alasaarela, and H. Lee, *"Secure Key Management Scheme based on ECC Algorithm for Patient's Medical Information in Healthcare System,"* *The International Conference on Information Networking (ICOIN2014)*, pp. 453–457, Feb. 2014. Phuket, Thailand: IEEE.

[39] S. T. Ali, V. Sivaraman, D. Ostry, G. Tsudik, and S. Jha, "Securing First-Hop Data Provenance for Bodyworn Devices Using Wireless Link Fingerprints," *IEEE Transactions on Information Forensics and Security*, Vol. 9, Issue 12, pp. 2193–2204, Dec. 2014.

[40] K. Singh, and L. Sharma, "Hierarchical Group Key Management Using Threshold Cryptography in Wireless Sensor Networks," *International Journal of Computer Applications*, Vol. 63, Issue 4, pp. 43–49, Feb. 2013.

[41] H. Wang, H. Fang, L. Xing, and M. Chen, *"An Integrated Biometric-based Security Framework Using Wavelet-domain HMM in Wireless Body Area Networks (WBAN),"* in *IEEE International Conference on Communications (ICC)*, pp. 1–5, June 2011. Kyoto, Japan: IEEE.

[42] L. Wang, Z. Lou, K. Jiang, and G. Shen, "Bio-Multifunctional Smart Wearable Sensors for Medical Devices," *Advanced Intelligence System*, Vol. 1, Issue 5, pp. 1–22, 2019.

[43] J. Singh, T. Pasquier, J. Bacon, H. Ko, and D. Eyers, "Twenty Security Considerations for Cloud-Supported Internet of Things," *IEEE Internet Of Things Journal*, Vol. 3, Issue 3, pp. 269–284, June 2016.

[44] M. Chiang, and T. Zhang, "Fog and IoT: An Overview of Research Opportunities," *IEEE Internet of Things Journal*, Vol. 3, Issue 6, pp. 854–864, Dec. 2016.

[45] A. John Stankovic, "Research Directions for the Internet of Things," *IEEE Internet of Things Journal*, Vol. 1, Issue 1, pp. 3–9, Feb. 2014.

[46] S. L. Keoh, S. S. Kumar, and H. Tschofenig, "Securing the Internet of Things: A Standardization Perspective," *IEEE Internet of Things Journal*, Vol. 1, Issue 3, pp. 265–275, June 2014.

[47] Y. Yang, L. Wu, G. Yin, L. Li, and H. Zhao, "A Survey on Security and Privacy Issues in Internet-of-Things," *IEEE Internet of Things Journal*, Vol. 4, Issue 5, pp. 1250–1258, Oct. 2017.

10 Access Control Mechanism in Health Care Information System

Bipin Kumar Rai
ABES Institute of Technology, India

Tanu Solanki
Galgotia College of Engineering and Technology, India

CONTENTS

10.1 Introduction 149
10.2 Access Control Mechanism... 151
    10.2.1 Discretionary Access Control (DAC) 151
    10.2.2 Mandatory Access Control (MAC)... 151
    10.2.3 Role-Based Access Control (RBAC).. 152
10.3 Access Control Solutions Associated with Health Care System................. 153
    10.3.1 Privacy-Aware Role-Based Access Control (P-RBAC)................. 153
    10.3.2 Personalized Access Control 153
    10.3.3 Context-Related Access Control.. 153
    10.3.4 Audit-Based Access Control.. 153
    10.3.5 Behavior-Based Access Control 153
    10.3.6 Rule-Based Access Control Approach... 153
    10.3.7 OASIS Role-Based Access Control.. 154
    10.3.8 XACML-Based Access Control 154
    10.3.9 Cryptography-Based Access Control... 155
10.4 Directions of Access Control Mechanism for Health Care Systems........... 156
    10.4.1 Process-Based Access Control 156
    10.4.2 Access Control for Patient-Controlled Electronic Health Records ... 157
10.5 Evaluation Criteria 157
10.6 Conclusion.. 159
References... 159

10.1 INTRODUCTION

In any organization, among the primary responsibilities of top management is to have satisfactory information security of information systems. The applications in organizations, which usually deal with safety and privacy, include access control [1–3].

DOI: 10.1201/9781003145042-10

Access control is a mechanism of information security to determine the resources or objects (access control list) allowed in an organization to legitimate subjects (users). Access control mediates every attempt of subjects or users in organizations to gain access to the control list (objects). Only after successful authentication and verification, comprehensive access is granted to verified users. The fundamental objectives of the access control in any information system are appropriately protecting the resources of system from inappropriate users and making sure the information is available to both users and applications [1,4,5]. Electronic health record (EHR) information systems need to develop a strong mechanism to protect unauthorized access to the data [6–9]. Obviously the nature and constraints of the particular information system affect the access control methodology, as well as its architecture [10–12].

Whenever we think about developing an access control mechanism for EHR, the primary requirement is that it should be able to satisfy the needs of all entities (i.e., patients, medical practitioners, medical authorities, etc.) [2,3,13]. Each entity needs to access certain fields of the EHR in order to carry out his/her job. These various entities need the ability to set specific access controls policies over the records. For health care information systems, the following privacy and security requirements are crucial [1]:

- Every entity in health care should have the right to decide on a security policy and enforce it within its domain.
- Health care providers should have the flexibility to arbitrarily define the security of a particular document, if so required [1,4].
- Patients should have the right to have control over their own health records, including whether or not to grant access to certain medical practitioners [4].
- Patients should have the ability to delegate control over their health records to someone else, under certain conditions (e.g., mental illness) [4].
- It should not be a complex task to manage access control policies.

An access control mechanism is intended to limit the actions or operations that a legitimate user can perform. Some basic terms are:

"**Object** is the entity of access control to contain about the information or resources or contains the receiving information. Usually access to object of control access means to access potentially the contained information or resources i.e. in database records or fields, files, directories, process and programs. Additionally processors, printers, nodes of networks, video displays and different components are the objects in system [1]."

"**Subject** is the entity of access control included of personnel, users, process and devices etc. the function of subject in system is to flow the information among objects [1]."

"**Operation:** a subject would like to invoke an active process is called operation of access control. E.g When users enter a card with correct PIN to ATM

Access Control Mechanism in Health Care Information System 151

machine, here the control program operation is considered a process on the user's behalf, numbers of several operations like deposit, balance inquiry and withdrawal are initiated by the subjects [1]."

"Privileges (Permission) is that kind of entity of access control to authorize some performance to some action in system [1]."

"Access Control List (ACL) contains all lists with possible objects in system and mechanism is to specify all subjects where they are able to access the objects in system along with possible initialized rights. In ACL there is a pair of subject with set of rights of each entry [1]."

"Access Control Matrix (ACM) it is a table of access control where a subject is represented by each row. Each column represents an object [1]."

10.2 ACCESS CONTROL MECHANISM

In the literature, three standard access control mechanism have been recognized. Each standard was designed to solve limitations found in a previous one [14].

1. Discretionary Access Control (**DAC**)
2. Mandatory Access Control (**MAC**)
3. Role-Based Access Control (**RBAC**)

10.2.1 DISCRETIONARY ACCESS CONTROL (DAC)

In a DAC mechanism, each user accesses his/her information on the basis of his identity and authorization. "DAC policies are commonly implemented through Access Control Lists (ACLs) and owner/other access control mechanisms" [4].

Limitations of DAC
- These mechanisms are difficult to manage, because the addition and deletion of users or data objects requires the discovery and treatment of all dependent entries in the DAC matrix.
- The DAC model might create new security problems, due to a patient's mismanagement of their own record [4].

10.2.2 MANDATORY ACCESS CONTROL (MAC)

The MAC mechanism provides access control on the basis of the security classification of subjects (users) and objects in the system.

Access control decisions are made by a central authority, not by the individual owner of an object, and the owner cannot change access rights. The need for a MAC mechanism arises when the security policy of a system dictates that protection decisions must not be decided by the object's owner, and the system must enforce data protection decisions [4].

As medical authorities must be responsible for assigning access rights to health care entity, some form of MAC policy should be involved.

Limitations of MAC

- It is very difficult, "due to the huge number of users who participate in EHR systems, the wide range of data types, and the desire to give patients ownership and partial control over their own medical records" [4].
- Limited scope because of unsuitability to patient control of their medical data.

10.2.3 ROLE-BASED ACCESS CONTROL (RBAC)

The RBAC mechanism has received a lot of attention in health care security research, due to its ability to provide a practical fine-grained access policy administration for a large number of users and resources. Most of the existing access control mechanisms for health care information system are either based on this mechanism or in some sense evolved from this mechanism [5].

RBAC controls access to information on the basis of the activities that particular types of users may execute in the system. Access decisions are taken based on the roles that individual users have as part of an organization. Users take on assigned roles (e.g., doctor, nurse, receptionist, etc.) [5].

Access rights (or permissions) are grouped by role name, and the use of resources is restricted to authorized individuals. The use of role hierarchies provides additional advantages, since one role may implicitly include the operations associated with another role.

In RBAC, access is granted according to rules expressed in the form of authorization policies (XACML) matching the role the current user embodies. RBAC can satisfy the least privilege access requirement, which involves granting the minimum set of privileges required for individuals to perform their job functions.

The separation of duty is incorporated into the RBAC mechanism to ensure that a user would not be allowed to execute two roles simultaneously, as per the organization's policy. RBAC is usually implemented where restricting network access based on the roles of individual users within an enterprise to access. Some of the available examples include Austria's ELGA (electronic health record) and the UK National Health Service

Limitation of RBAC

- A weakness of this mechanism is the central access control module, which can be bypassed or circumvented by administrators with unrestricted access rights.
- Some evaluations of access requests are complex, as it needs to consider contextual parameters.
- It is not sufficient for the privacy and security requirements of health care systems.

We found that "none of DAC, MAC, RBAC mechanisms is sufficient for the privacy and security requirements of Electronic Health Record systems. However a careful combination of all three access control standards can be used to deliver the essential security requirements" [14].

10.3 ACCESS CONTROL SOLUTIONS ASSOCIATED WITH HEALTH CARE SYSTEM

10.3.1 PRIVACY-AWARE ROLE-BASED ACCESS CONTROL (P-RBAC)—[14]

P-RBAC extends RBAC to provide full support for expressing highly complex privacy related policies having features like obligations and purposes. Privacy policies are expressed as permission assignments, which is different from RBAC. It is similar to XACML and directly supports fine-grained policies by its native support of conditions/obligations.

Qun et al. [15] present a "flexible authoring tool based on the SPARCLE system supporting the high level specification of P-RBAC Permissions."

10.3.2 PERSONALIZED ACCESS CONTROL

This combines the concepts of RBAC and DAC. In this approach, an owner (patient) will decide who can access her/his health data.

10.3.3 CONTEXT-RELATED ACCESS CONTROL

For health care—
This method is applied to a health care business process control in contextual parameters (e.g., time, location, etc.) to RBAC that involves multiple actors accessing the data and resources needed for performing clinical and logistics tasks in the application [14,16].

10.3.4 AUDIT-BASED ACCESS CONTROL

Audit logic is an a posteriori access control framework useful for urgent situations. It allows medical officers to handle exceptional situations and justify it later. Here "a-priori access control is minimized to authentication of users and objects and the basic authorizations [14], [17].

10.3.5 BEHAVIOR-BASED ACCESS CONTROL

This is for making the model compatible with different health care environments [18].

10.3.6 RULE-BASED ACCESS CONTROL APPROACH

This is for a resource-limited pervasive health care system. "It allows users to access systems and information based on pre-determined and configured rules. But there is no commonly understood definition or formally defined standard for rule-based access control as for DAC/MAC/RBAC. It can be

154 Cybersecurity

combined with other models particularly RBAC or DAC. A Rule based access control system intercepts every access request and compares the rules with the rights of the user to make an access decision. Its rules cannot be changed by users. The rules can be created by any attributes of a system related to the users such as domain, host, network, IP addresses, and protocol" [1].

10.3.7 OASIS ROLE-BASED ACCESS CONTROL

Developed prototype system including a basic network and role-based security infrastructure for the United Kingdom National Health Service. Cassandra, a role-based policy specification language for expressing access control policies in large-scale distributed systems, is used to specify security policies for the UK national EHR system. By reducing administrative complexity, role-based access control has become the common theme applied in these approaches. However, the EHR considered in these approaches is either a general abstract object or an isolated primitive object [19].

10.3.8 XACML-BASED ACCESS CONTROL

Figure 10.1 shows the sequence of the following steps-

1. The policy administration point is responsible for writing the policies and policy sets which represent the complete policy for a specified target and make them available to the policy decision point (PDP).
2. The requester sends a request for access to the policy enforcement point (PEP).
3. The PEP asks the context handler for access in its native request format, which optionally includes attributes of the subjects, resource, action, environment, and other categories.
4. Then context handler creates an XACML request context, optionally adds attributes, and sends it to the PDP.
5. After that, the PDP requests any additional subject, resource, action, environment, and other categories (not shown) attributes from the context handler.
6. The context handler requests the attributes from a policy information point (PIP).
7. The PIP obtains the requested attributes.
8. The PIP returns the requested attributes to the context handler. The context handler includes the resource in the context (optionally).
9. The context handler sends the resource (optionally) to the PDP.
10. The context handler sends the requested attributes to the PDP. Then, the PDP evaluates the policy.
11. After evaluation, the PDP returns the response context (including the authorization decision) to the context handler.
12. The context handler converts the response context to the native response format of the PEP and then returns the response to the PEP.
13. Then PEP fulfills the obligations. The PEP permits access to the resource, if access is permitted; otherwise, it denies access [20].

Access Control Mechanism in Health Care Information System

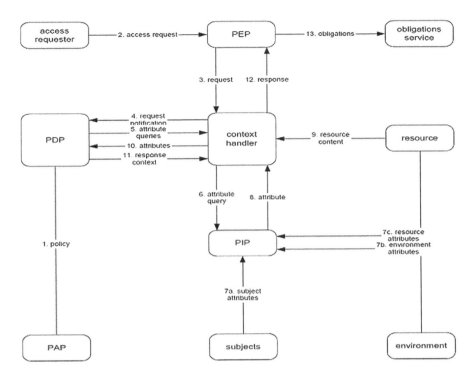

FIGURE 10.1 Data flow of XACML-based access control [20].

10.3.9 Cryptography-Based Access Control [14]

This is a new access control methodology that defines an implicit access mechanism that uses cryptography to provide security and privacy in health care. It is a logical control function for sharing resources and accessing appropriate objects.

The basic idea is to provide a unique key (class key) for each resource type, so that users belonging to a class are assigned the key for that class.

A key is provided by the user to access that resource which is verified against the resource key. If they match, then access is granted; otherwise, access is denied.

In hierarchical access control, parent class users can access the resources of the child classes.

This does not require an access control matrix for storing accessibility information for each object and subject and hence is suitable for a web-based distributed environment.

It eliminates the need for a reference monitor (which is an operating system component that validates user requests for access to resources against an access control scheme) and relies exclusively on cryptography.

Data are encrypted before storage on server. A separate key pair may be connected with every file and should only be distributed to the main entity allowed to access that file. Data are encrypted with a private key (also correspond to a write authority) and decrypted with a public key (also corresponds to a read authority). The access control mechanism decides who has permission to access; semantics are interpreted by the object server only.

For better performance, an enhanced cryptography access control mechanism suggests symmetric encryption for encrypting file and then asymmetric encryption to sign the file to ensure integrity.

Limitation of cryptography-based access control
* A comparatively new mechanism, it needs to be standardized, tested, and published
* A large number of keys are needed and hence must be managed without revealing the privacy of the owner.
* Processing speed is a big issue, as asymmetric key encryption is used based on the RSA family and a large number of prime numbers are used. One solution could be elliptic curve cryptography based an access control mechanism.

10.4 DIRECTIONS OF ACCESS CONTROL MECHANISM FOR HEALTH CARE SYSTEMS

Access control mechanisms need be developed to achieve the following:

* To collect knowledge that forms a foundation for access control requirements in health care systems.
* To create improved access control mechanisms for health care systems, based on real requirements.

There are two main directions:

10.4.1 PROCESS-BASED ACCESS CONTROL

The requirements for health care suggest using knowledge in the form of observations, medical guidelines, and the mining of audit logs in access control. After that extraction, represent and combine this knowledge.

Extracting usage patterns: It is a nontrivial task to extract usage patterns by data mining of logs, which requires the availability of sufficiently large amounts of high-quality log data. Usage patterns may be simple or more complex and may include information on the location, responsibility, and roles of the present users, the time, and the situation.

Rules for creating permissions: If a usage pattern is common, that does not necessarily mean that it should be included as an access rule. In same way, because an event is uncommon does not necessarily mean that it should be disallowed.

Misuse detection: Usage patterns may also be used to create application-level intrusion detection systems (IDS) for health care that help automate the process of misuse detection.

Access Control Mechanism in Health Care Information System   157

10.4.2 Access Control for Patient-Controlled Electronic Health Records

A correctly implemented mechanism is not sufficient for access control in a process-controlled EHR. Further refinement should focus on:

Usability: As the patient becomes the administrator, the usability of the systems in the way it communicates consequences to the user is important.

Common policies: If the system suggests an access profile, it seems likely that users will trust the system and accept that suggestion without closer examination.

Creating common policies based on the most common personal policies: Granting users the power to create their own access profiles means that over time, we get a collection of access profiles representing who users share with and what they share.

10.5 EVALUATION CRITERIA

Eric Helms identifies 25 evaluation criteria to evaluate the RBAC implementation of EHRs.

Three come from the HIPAA Security Rule and represent legal requirements of the EHR. Eight come from certification commission for health information technology (CCHIT) and national institute of standards and technology (NIST) Meaningful Use that represent compliance criteria for ARRA incentives.

Eight are from the NIST RBAC standard and represent best practices in RBAC implementation. The final six are best practices drawn from related works that discuss additions to standard RBAC to address security concerns.

Each criterion is given a classification that adheres to the convention of an acronym followed by a number index. The classifications are: (HIPAA), CCHIT Meaningful Use (CCHIT), NIST Meaningful Use (NIST-MU), NIST RBAC standard best practice (NIST-BP), and literature best practices (L-BP) [21].

HIPAA Criteria

The HIPAA Security Rule defines three requirements for access control. First is to use a unique identifier for each user; second is to have general access control present within the system; and third is to have emergency access procedures to gain access to needed electronic protected health information in emergency situation if normal systems operations have been damaged.

HIPAA-1: Unique User Identifier

HIPAA-2: General Access Control

HIPAA-3: Emergency Access Procedures [21]

CCHIT Criteria

CCHIT-1: Users given least privilege permission set

CCHIT-2: Administrative ability to assign privileges to users/groups

CCHIT-3: The system must implement one of user-based, role-based, or context-based access controls

CCIHT-4: The system should allow a user to remove his permissions without deleting the user [21]

NIST Meaningful Use

HIPAA defines a similar requirement for a unique user identifier (HIPAA-1), but the NIST standard is specific to name and/or number. NIST-MU-2 and HIPAA-2 are similar requirements, but the NIST criteria specifically mention user privileges, whereas HIPAA simply states the use of access control.

NIST-MU-1. User-Permission Assignment

NIST-MU-2. Emergency Access Roles/Permissions [21]

The last two requirements are for emergency situations when user is authorized for a set of privileges that are only applicable during an emergency situation and that there should exist the ability to activate emergency access role usage.

NIST-MU-3. Emergency Role Activation

NIST-MU-4. User identified by unique name and/or number

NIST Flat RBAC

NIST-BP-1: Users

NIST-BP-2: Roles

NIST-BP-3: Permissions

NIST-BP-4: Multiple roles per user

NIST-BP-5: User-Role Review

NIST Hierarchical RBAC

Hierarchy in roles allows inheriting permissions from multiple roles in the general case and from a single role in the limited case

NIST-BP-6: Role Hierarchies

NIST Constrained RBAC

This incorporates the policy of separation of duty to prevent a single user from performing actions that may provide a conflict of interest.

NIST-BP-7: Separation of Duty

NIST Symmetric RBAC [21]

This is built upon the previous three levels. "Similar to Flat RBAC's requirement for user-role review, it establishes a requirement for an implementation to be able to have its permission-role relationships reviewed, **in addition to** NIST-BP-8: Permission-Role Review."

Other Criteria [21]

L-BP-1: Administrative Facilities

L-BP-2: Lack of Super User

L-BP-3: Role Creation and Assignment Separation of Duty

L-BP-4: Static or Dynamic Role Definitions

L-BP-5: Permissions: Positive, Negative, or Both

L-BP-6: RBAC Core Application or Module

10.6 CONCLUSION

Every Information system has different needs and requirements, according to the nature of the domain. A correctly implemented mechanism is not sufficient for access control in a process-controlled EHR. So, balance is required to increase the acceptance of clinicians, confidence, and trust of the patient. However, the patient-oriented mechanism for the health care information system is more appealing, as it ensures the privacy rights of patients at any cost.

REFERENCES

1. Alhaqbani, Bandar, and Colin Fidge. *"Access control requirements for processing electronic health records."* In *International Conference on Business Process Management*, pp. 371–382. Berlin, Heidelberg: Springer, 2007.
2. Rai, Bipin Kumar, and A. K. Srivastava. "Security and privacy issues in healthcare information system." *International Journal of Emerging Trends & Technology in Computer Science (IJETTCS)* ISSN 2278-6856: 3.6 (2014): 248–252.
3. Rai, Bipin Kumar, and A. K. Srivastava. "Pseudonymization techniques for providing privacy and security in EHR." *International Journal of Emerging Trends & Technology in Computer Science (IJETTCS)* ISSN 2278-6856: 5.4 (2016): 34–38.
4. Røstad, Lillian. *Access Control in Healthcare Information Systems*, pp. 1–158. PhD Dissertation, NTNU Trykk, (2008).
5. Rai, Bipin Kumar, *Pseudonymization Based Mechanism for Security & Privacy of Healthcare* in 2020. Latvia: Lambert Academic Publishing.
6. Rai, Bipin Kumar, and A. K. Srivastava. "Patient controlled pseudonym-based mechanism suitable for privacy and security of Electronic Health Record." *International Journal of Research in Engineering, IT and Social Sciences, (IJREISS)* ISSN 2250-588: 7.2 (2017): 26–30.
7. Rai, Bipin Kumar, and A. K. Srivastava. "Prototype implementation of patient controlled pseudonym-based mechanism for Electronic Health Record (PcPbEHR)." *International Journal of Research in Engineering, IT and Social Sciences (IJREISS)* ISSN 2250-588: 7.7 (2017): 22–27.
8. Frikken, Keith, Mikhail Atallah, and Jiangtao Li. "Attribute-based access control with hidden policies and hidden credentials." *IEEE Transactions on Computers* 55.10 (2006): 1259–1270.
9. Gajanayake, Randike, Renato Iannella, and Tony R. Sahama. *"Privacy oriented access control for electronic health records."* In *Data Usage Management on the Web Workshop, at the Worldwide Web Conference*, pp. 9–16. Germany: ACM, 2012.
10. Bacelar-Silva, Gustavo Marísio, Margarida David, Luís Antunes. *"Comparing security and privacy issues of EHR: Portugal, the Netherlands and the United Kingdom."* In *Proceedings of the 4th International Symposium on Applied Sciences in Biomedical and Communication Technologies*, pp. 1–4. Barcelona, Spain: ACM, 2011.

11. Hue, Pham Thi Bach, Sven Wohlgemuth, Isao Echizen, Nguyen Dinh Thuc, Dong Thi Bich Thuy. "An experimental evaluation for a new column-level access control mechanism for electronic health record systems." *International Journal of u-and e-Service, Science and Technology* 4.3 (2011): 73–86.

12. Zheng, Yi, Yongming Chen, and Patrick CK Hung. "*Privacy access control model with location constraints for XML services.*" In *2007 IEEE 23rd International Conference on Data Engineering Workshop*, pp. 371–378. Istanbul, Turkey: IEEE, 2007.

13. Jin, Jing, Gail-Joon Ahn, Michael J. Covington, and Xinwen Zhang. "*Access control model for sharing composite electronic health records.*" In *International Conference on Collaborative Computing: Networking, Applications and Worksharing*, pp. 340–354. Berlin, Heidelberg: Springer, 2008.

14. Al-Hamdani, Wasim A. "*Cryptography based access control in healthcare web systems.*" In *2010 Information Security Curriculum Development Conference*, pp. 66–79. Kennesaw, Georgia: ACM, 2010.

15. Ni, Qun, et al. "Privacy-aware role-based access control." In *ACM Transactions on Information and System Security (TISSEC)* 13.3 (2010): 24.

16. Kulkarni, Devdatta, and Anand Tripathi. "*Context-aware role-based access control in pervasive computing systems.*" In *Proceedings of the 13th ACM Symposium on Access Control Models and Technologies*, pp. 113–122. Estes Park, CO, USA: ACM, 2008.

17. Dekker, Mari Antonius Cornelis, and Sandro Etalle. "Audit-based access control for electronic health records." *Electronic Notes in Theoretical Computer Science* 168 (2007): 221–236.

18. Yarmand, Mohammad H., Kamran Sartipi, and Douglas G. Down. "*Behavior-based access control for distributed healthcare environment.*" In *2008 21st IEEE International Symposium on Computer-Based Medical Systems*, pp. 126–131. Jyvaskyla, Finland: IEEE, 2008.

19. Eyers, David M., Jean Bacon, and Ken Moody. "OASIS role-based access control for electronic health records." *IEE Proceedings-Software* 153.1 (2006): 16–23.

20. http://docs.oasis-open.org/xacml/3.0/xacml-3.0-core-spec-cs-01-en.pdf, Nov. 19, 2020.

21. Helms, Eric, and Laurie Williams. "*Evaluating access control of open source electronic health record systems.*" In *Proceedings of the 3rd Workshop on Software Engineering in Health Care*, pp. 63–70. Honolulu: ACM, 2011.

11 Privacy Preservation Tools and Techniques in Artificial Intelligence

Raneem Qaddoura and Nameer N. El-Emam
Philadelphia University, Jordan

CONTENTS

11.1 Introduction 161
11.2 Related Work 162
11.3 Particle Swarm Optimization (PSO) 163
11.4 Datasets 164
11.5 Evaluation Measures 165
11.6 Machine Learning Techniques with PSO 167
    11.6.1 Support Vector Machine (SVM) 167
    11.6.2 Random Forest (RF) 170
    11.6.3 Neural Network (NN) 171
    11.6.4 k-Nearest Neighbor (k-NN) 174
    11.6.5 Other Related Techniques Available 174
    11.6.6 Discussion and Recommendations 175
11.7 Conclusion and Future Work 176
References 176

11.1 INTRODUCTION

Network security has become a substantial area in today's life. Data needs to be carefully protected, as it has become a very powerful asset and a success factor for many industries [3]. This introduces the need for intrusion detection systems to monitor networks and analyze their activities for possible attacks, which prevent unauthorized users from accessing these networks [4]. Attackers try to exploit the vulnerability of the networks to gain illegal access to nonpublic data and resources, which results in the disruption of the regular activities of the networks [4].

Many approaches have been proposed to automatically detect the attacks caused by unauthorized access. Ongoing work on this is still desirable, as no optimal technique has been found yet, and attackers are continuously changing their way of gaining illegal access to networks [5].

Recent intrusion detection systems use machine learning classification techniques to detect possible attacks. These techniques try to increase the detection rate

DOI: 10.1201/9781003145042-11

161

162   Cybersecurity

and decrease the false alarms of predicted attacks [4]. These challenges sometimes exist due to the redundancy and irrelevancy of features, which are considered while training and generating the classification model [5]. Other causes for a low detection rate and high false alarm rates include the choice of the classification technique and the inaccurate tuning of the parameters needed for using these techniques [6]. To cope with these problems, optimization techniques are used to select the datasets' features and the techniques' parameters while running the selected classification technique [5]. The particle swarm optimization (PSO) technique is widely used in the literature to optimize the classification techniques in solving the intrusion detection problem.

This chapter discusses different classification techniques combined with PSO for intrusion detection. Publications are gathered considering PSO and intrusion detection for the duration between 2015 and 2020. Further manual affirmation of the relevancy of the publications is performed. The chapter covers the following topics:

1. The introduction of different PSO-based techniques used in intrusion detection.
2. Discussion of datasets used in the literature for intrusion detection.
3. Presentation of the most common measures used to evaluate intrusion detection systems.
4. Insights into and directions for the use of different techniques for intrusion detection.

The remainder of this chapter is organized as follows: Section 11.2 surveys the related work found in the literature. Section 11.3 provides an overview of the PSO algorithm. Sections 11.4 and 11.5 discuss the datasets used by most works, and the metrics used for evaluation, respectively. Section 11.6 discusses in detail different classification techniques used with PSO for intrusion detection. It provides comparisons between different techniques and recommendations for selecting these techniques. Finally, the last section concludes the work and provides future possible directions for study.

11.2 RELATED WORK

PSO can be found extensively in the literature to solve different problems such as steganography [7], medical diagnosis [8], email spam filtering [9], botnet detection in the Internet of Things [11], and smart cities [12]. Z. Yudong et al. (2015) [13] present a comprehensive survey on PSO, including the modifications advanced on PSO, population topology, hybridization, extensions, theoretical analysis, parallel implementation, and applications. A similar survey is presented in [14], which describes the variations of the algorithm, and the modifications and refinements of the hybridization of PSO with other heuristic algorithms. The work presented in [15] emphasizes the development, deployment, and advancement implementations. It provides recommendations for inertia weight, constriction factor, cognition, and social weights selection. Other surveys are specific to multimodal PSO [16], multi-objective PSO [17], parallel PSO [18], and clustering with PSO [19–24].

Privacy Preservation Tools and Techniques

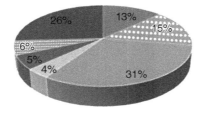

■ SVM □ RF ■ NN ■ SMO ■ Knn ▩ MCLP■ others

FIGURE 11.1 Percentage of publications with different techniques.

Some other surveys are specialized in their review by specific applications. PSO can be found in the applications of the antenna circuit [25], stock market [26], medical data clustering [27], university examination timetabling problem [28], engineering and network [29], and wireless sensor networks [30].

Another important application is intrusion detection, which is widely covered in the literature. Many classification techniques in intrusion detection have been proposed combining PSO with other techniques, such as support vector machine (SVM) [6,31,32], random forest (RF) [33–36], neural network (NN) [5,37–39], k-nearest neighbor (k-NN) [40,41], multiple-criteria linear programming

(MCLP) [42,43], sequential minimal optimization (SMO) [44], and others [44–46,39]. Figure 11.1 shows the percentage of publications using different techniques. It is observed from the figure that a large portion of techniques combine PSO with neural network, and a good portion of techniques consider SVM and RF with PSO, while others include k-NN, MCLP, and other techniques.

For this reason, there is a persistent need to review all techniques in a survey, to summarize and compare different techniques, and provide recommendations and insights about possible enhancements. To the best of our knowledge, no survey covers all the classification techniques combined with PSO for intrusion detection. Thus, this survey covers this area by discussing different PSO-based techniques for intrusion detection and providing insights and directions for the use of these techniques for intrusion detection.

11.3 PARTICLE SWARM OPTIMIZATION (PSO)

Particle swarm optimization is a nature-inspired optimization algorithm that is inspired by a flock of birds, a school of swimming fish, or a colony of ants [43]. It is based on a population of particles found by J. Kennedy and R. Eberhart (1995) [48]. The algorithm starts by initializing a population in space with random positions and velocities. The algorithm aims to find a better position for particles by updating the particles' positions toward an optimal solution based on the values returned by a fitness function [39]. For each population in an iteration, particles move toward the optimal solution by equations 11.1 and 11.2, for an ith particle in a population [41]:

$$v_{i+1}^i = w.v_t^i + c_1 r_1 \left(p_t^{i,best} - p_t^i \right) + c_2 r_2 \left(p_t^{g,best} - p_t^i \right) \quad (11.1)$$

$$p_{t+1}^i = p_t^i + v_{t+1}^i \tag{11.2}$$

where t is the current time, v is the velocity, p is the position, $p^{i,best}{}_t$ is the best own position of previous populations, $p^{g,best}{}_t$ is the best position for all particles in previous populations, c and r are positive acceleration coefficients and d-dimensional vectors of random numbers, respectively. Finally, w controls the pressure of local and global searches.

Optimization algorithms aim to search the space to find an optimal or near-optimal solution [49]. PSO is used along with classification techniques to solve the intrusion detection problem. PSO has a simple implementation and is scalable, robust, and flexible. It converges relatively faster than other nature-inspired optimization algorithms and gives good quality results [41].

11.4 DATASETS

Among the publications gathered for this study, there is an agreement on the dataset used in the experiments; almost all publications used either the KDD99, NSL-KDD, or UNSW-NB15 datasets.

The KDD99 dataset is publicity available and is widely used in the literature to evaluate intrusion detection. It has 41 features, which are detailed in [50], and about five million instances where each instance is labeled as either normal or intrusive. According to the work presented in [32], features are classified as basic, content, traffic, and host traffic.

The KDD99 dataset faces different types of attacks: denial-of-service (DoS), surveillance or probe (Probe), user to root (U2R), and remote to local (R2L) [32,50]. DoS refers to preventing permitted users from accessing a service. Probe refers to detecting weaknesses in the machine. R2L refers to gaining illegal access, and U2R refers to gaining super-permissions while gaining access [41].

On the other hand, the NSL-KDD is an improved version of the KDD99 that resolves some of the issues observed in KDD99. These issues include redundant instances that cause bias by the learning algorithm for repeated instances. Also, high accuracy scores are observed by typical learning algorithms, which makes evaluating the accuracy of the proposed methods inappropriate [51]. The NSL-KDD dataset is introduced to resolve such issues; redundant instances are removed, and the number of instances for each difficulty level group and the number of instances for the training and testing parts are optimized, which makes the evaluation consistent and comparable for different works.

In addition, the UNSW-NB15 dataset [52] includes nine attack types, 49 features, and 2,540,044 instances. The dataset's features are classified into six different categories: basic, flow, time, content, generated, and labelled features.

Figure 11.2 shows the percentage of publications using these datasets. It is observed from the figure that almost half of publications use the KDD99 dataset, and a large portion of publications are experimenting with the NSL-KDD dataset. In addition, a few publications are using the UNSW-NB15 dataset. Figure 11.3 shows

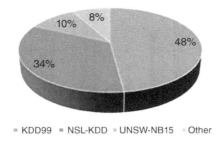

FIGURE 11.2 Percentage of publications per dataset.

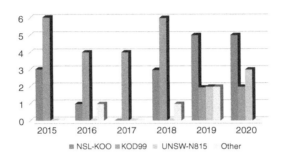

FIGURE 11.3 Number of publications per year and dataset.

the distribution of publications using the datasets, by publication year. It is also observed from Figure 11.3 that the use of the NSL-KDD and UNSWNB15 datasets relatively increases for the years 2019 and 2020, as the recognition by researchers for the existence of the UNSW-NB15 dataset and the enhancements made on the old version of the NSL-KDD dataset are observed [53–55].

11.5 EVALUATION MEASURES

In the literature, the most common measures for evaluating the efficiency and performance of the proposed techniques include accuracy (ACC), the error rate (ER), the true positive rate (TPR), the false positive rate (FPR), the true negative rate (TNR), the false negative rate (FNR), precision (PREC), the F1-score (F1), the correlation coefficient (CC), and detection time. Figure 11.4 shows the number of publications by each evaluation measure. It is observed from the figure that the ACC, TPR, and FPR are extensively considered in the literature.

The performance of a classification algorithm is measured based on how well it recognizes attacks from the normal cyberattacks. The number of correct predictions of attacks is referred to as the true positive (TP), while the number of correct predictions of normal labels is referred to as the true negative (TN). In contrast, the number of wrong predictions of attacks is referred to as the false Positive (FP), while the number of the wrong predictions of normal labels is referred to as the

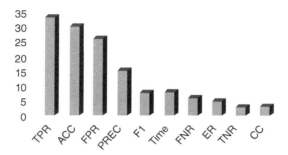

FIGURE 11.4 Number of publications per evaluation measure.

false negative (FN). In this sense, the description of the evaluation measures is as follows [5,32]:

- ACC: The ratio of the number of correct classifications of normal and attack labels to the number of all instances. It is calculated by:

$$ACC = \frac{TP+TN}{TP+TN+FP+FN} \quad (11.3)$$

- ER: The ratio of the number of wrong classifications of normal and attack labels to the number of all instances. It is calculated by:

$$ER = \frac{FP+FN}{TP+TN+FP+FN} \quad (11.4)$$

- TPR: Also called the detection rate (DT), recall (REC), or sensitivity. It is the ratio of the number of correct classifications of attack labels to the number of predicted attack labels. It is calculated by:

$$TPR = \frac{TP}{TP+FN} \quad (11.5)$$

- FPR: Also called the false alarm rate (FAR). It is the ratio of the number of wrong classifications of attacks to the number of predicted normal labels. It is calculated by:

$$FPR = \frac{FP}{TN+FP} \quad (11.6)$$

- FNR: Also called the miss rate. It is the ratio of the number of wrong classifications of normal labels to the number of predicted attack labels. It is calculated by:

$$FNR = \frac{FN}{TP+FN} \quad (11.7)$$

Privacy Preservation Tools and Techniques 167

- TNR: Also called specificity. It is the ratio of the number of correct classifications of normal labels to the number of predicted normal labels. It is calculated by:

$$TNR = \frac{TN}{TN+FP} \tag{11.8}$$

- PREC: The ratio of the number of correct classifications of attack labels to the number of actual attack labels. It is calculated by:

$$PREC = \frac{TP}{TP+FP} \tag{11.9}$$

- F1: The harmonic mean of precision and recall (TPR).

$$F1 = \frac{2 \times PREC \times REC}{PREC+REC} \tag{11.10}$$

- CC: A balanced measure that takes into consideration the correlation between the predicted and actual labels. It is calculated by:

$$CC = \frac{TP \times TN - FP \times FN}{\sqrt{(TP+FN) \times (TP+FP) \times (TN+FP) \times (TN+FN)}} \tag{11.11}$$

11.6 MACHINE LEARNING TECHNIQUES WITH PSO

This section discusses in detail different machine learning (ML) techniques used with PSO for intrusion detection. The techniques are compared, with recommendations and directions for using them in intrusion detection.

Table 11.1 shows how different combinations of PSO and other techniques are found in the publications considered for this work. As observed from the table, SVM, RF, NN, SMO, k-NN, MCLP, and other techniques are optimized by PSO for solving the intrusion detection problem. It is also observed that NN, SVM, and RF are the most used techniques in combination with PSO for intrusion detection. In addition, some publications combine more than one technique to detect intrusion.

11.6.1 SUPPORT VECTOR MACHINE (SVM)

The main goal of SVM is to find a hyperplane that increases the distance between the separating hyperplane and the closest data points of the input data for each class [43]. If data is not linearly separable, a kernel function is used to extend the linear classification ability to an extended nonlinear one [72]. SVM has a low generalization error rate, so it does not suffer much from overfitting to the training dataset [41].

TABLE 11.1
PSO-based techniques for intrusion detection

Ref.	SVM	RF	NN	k-NN	SMO	MCLP	PCA	Xgboost	DT	Others
[3]			x							
[4]			x							
[5]			x							
[6]	X									
[31]	X									
[32]	X									
[33]		X	X							
[34]		X			X					
[35]		X	X							
[36]		X								
[37]			X							
[38]			X							
[39]			X							
[40]				X						
[41]	X			X						
[42]						X				
[43]	X					X				
[44]					X					
[45]								X		
[46]										X
[47]										X
[53]		X								X
[54]			X							
[55]			X							
[56]			X							
[57]										X
[58]								X		
[59]			X	X						
[60]										X
[61]										X
[62]									X	
[63]			X							
[64]									X	
[65]			X							
[66]										X
[67]			X							
[68]		X								
[69]										X
[70]		X								
[71]							X			
[72]	X						X			
[73]	X									
[74]										X
[75]						X				
[76]		X								
[77]			X							
[78]			X							

Privacy Preservation Tools and Techniques

SVM is extensively used for intrusion detection to determine the existence of attacks and specify types of attacks. However, there are two main problems with using SVM in intrusion detection:

- Choosing the optimal values of parameters for SVM, which are c, γ or σ, and ε.
- Selecting the optimal subset of features of the dataset.

These problems are addressed in many works where PSO is used for parameter selection and other methods are used for feature selection.

D. J. Kalita et al. (2020) [6] propose a technique named Multi-PSO-SVM, which optimizes the values of c and γ parameters in SVM by multi-PSO, which is a variant of PSO that first partitions the search space for the c and γ values and then deploys different swarms for each of the partitions. W. GuiPing (2015) [73] proposes multiclass-SVM-PSO (MSVM-PSO) by optimizing the values of c and γ parameters of a multiclass SVM by PSO, which not only specifies the existence of an attack but also specifies the type of attack.

The works implemented by Bin Tan et al. (2016) [32] and Ye Bi [31] propose improved PSO-SVM (IPSO-SVM) and hybrid PSO-SVM (HPSO-SVM) techniques, respectively. They consider optimizing the radial basis function (RBF) kernel parameter σ, penalty parameter c, and insensitive loss error ε parameters of SVM using PSO to detect intrusion. No feature selection techniques are used in these works to reduce the number of features, but the proposed approach in [32] considers 22 attributes that are the basic attribute set and the content attribute set of the KDD99 dataset, which are related to DoS and surveillance or probe attack types, without actually applying a technique for feature selection.

An intrusion detection ensemble construction method was proposed in [41] with six SVM experts, six k-NN experts and three ensembles using PSO, meta-optimized PSO based on local unimodal sampling (LUS), and a weighted majority algorithm (WMA). The ensembles arrive at a final decision by incorporating the opinions of the 12 experts. PSO is used to optimize the SVM RBF parameter γ and the k-NN k parameter.

In [72], a model of a combination of multilayered SVM, KPCA, and improved chaotic PSO (ICPSO) is proposed for intrusion detection. ICPSO optimizes the c, σ, and ε parameters of SVM, while KPCA and N-RBF are used for feature extraction and noise reduction, respectively.

Other works consider optimizing the selection of both features and SVM parameters' values. An effective framework for detecting intrusion is proposed by [43]. The framework considers a time-varying chaos particle swarm optimization (TVCPSO) with multiple criteria linear programming (MCLP) and SVM. The search engine of PSO contains the c and γ in SVM, and α and β in MCLP in addition to the binary features mask of the 41 feature sets.

Table 11.2 summarizes the techniques used, the parameters selected for optimization, the algorithms with which the results are compared, and the evaluation measures used to compare each technique. As observed from the table, different variants

TABLE 11.2
SVM based PSO techniques

Ref.	Technique	Param.	Compared with**	Measures
[6]	Multi-PSO-SVM	c,γ	Single-PSO, GS, GA	PREC-REC rate
[31]	HPSO–SVM	σ,c,ε	Single-SVM	Time, FPR, TPR, CC
[32]	IPSO-SVM	σ,c,ε	Single-SVM, NN, Adaboost, GA, SA, BO	ACC, TPR, FPR, PREC
[41]	SVM-kNN-PSO	γ	Single-SVM, NN, LUS, WMA	ACC
[43]	TVCPSO-MCLP	c,γ	MCLP, Single-SVM, C5.0	TPR, FPR, Time
[72]	KPCA-ICPSO-SVM	σ,c,ε	KPCA-CPSO-SVM, N-KPCAGA-SVM, KPCA-GA-SVM, PCA-GA-SVM, PCA-PSO-SVM, CPSO-SVM, Single-SVM, RBFNN	Time, ACC, TPR,CC
[73]	MSVM-PSO	c,γ	Xgboost, GS	TPR, PREC

** GS: grid search, GA: genetic algorithm, NN: neural network/artificial NN/deep NN SA: simulated annealing, BO: Bayesian optimization

of combining SVM and PSO are proposed and different parameters are optimized through PSO, which are: c, γ, σ, and ε parameters. The techniques are compared with different algorithms using different evaluation measures.

11.6.2 RANDOM FOREST (RF)

Random forest incorporates the decision tree and ensemble learning [33]. The collection of trees with an arbitrary subgroup of the data is formed by a bagging approach. The first phase of the algorithm initiates the trees and the second phase combines the trees that have the same test features. The algorithm then chooses the class that is classified by most trees as the predicted solution [68].

A combination of RF and PSO for intrusion detection is found in the literature in two ways: The first is when RF is applied for feature selection (FS) and PSO is used to optimize neural network parameters. The other applies PSO for feature selection and RF for intrusion classification.

The works of [33,35] apply RF for feature selection. L. Hengxun (2018) [35] proposes a hybrid feature selection model based on random forest and particle swarm optimization in which a weighted RF algorithm is used for feature selection and a two-layer neural network is modeled for evaluating the selected subset of features. A PSO algorithm is used to optimize the parameters building the RF. Feature selection is performed by RF in the proposed approach in [33] to eliminate irrelevant attributes, and neural network is used for classification. PSO is then applied on the selected features of the dataset to optimize the work of neural network by considering the forward propagation as an objective function.

Other works consider PSO for feature selection and RF for classifying data to detect intrusions. In [36], B. A. Tama and K. H. Rhee (2018) use this approach for anomaly detection in an IoT network and the grid search to determine the best

Privacy Preservation Tools and Techniques

TABLE 11.3
RF-based PSO techniques

Ref.	Approach	Compared with**	Measures
[33]	PSO to optimize NN parameters	k-NN, SVM, LR, DT, NB, PSO-discretize-HNB [79], MBGWO [80], MI-BGSA [81], DO-IDS [82]	ACC, PREC, F1, FPR, FNR, TPR, TNR
[34]	PSO for random values of ROC	Firefly, 544 Coral reef, harmonic search	ACC, ER, FPR, TPR
[35]	PSO to optimize NN parameters	CFS and SVM	TPR, FPR
[36]	PSO for FS	RoF, DNN	ACC, PREC, TPR, FPR
[68]	PSO for FS	single-RF, NB, PART, Bagging, Jrip, BayesNet, CART-BN [84]	TPR, FPR
[70]	Multi-objective PSO+RF+IG	N/A	ACC
[76]	PSO for FS	Bagging-C4.5, Real AdaboostC4.5, Multiboost-C4.5, Rotation Forest-C4.5, Maj. Voting (C4.5+RF+CART)	ACC,FPR
[53]	PSO for FS	DT, NB, logistic regression, NN, Expectation-Maximization	ACC, PREC, TPR, FPR

** LR: logistic regression, DT: decision tree, NB: naive Bayes, CFS: correlation-based feature selection, RoF: rotation forest, DNN: deep neural network

learning parameters for RF. The proposed approach in [68] uses different dimension reduction techniques and considers binary PSO for further dimensional reduction, then RF is applied as a classifier. T. Bayu Adhi and R. Kyung-Hyune (2015) [76] use an ensemble of tree-based classifiers (C4.5, RF, and CART) to classify the dataset with a reduced number of attributes specified by PSO. B. A. Tama et al. (2019) [53] apply a two-stage classifier ensemble (TSE-IDS) with feature selection using PSO, GA, and an ant colony algorithm on NSL-KDD and UNSW-NB15 datasets.

In the approach proposed by [34], RF is considered, along with naive Bayes, J48, logistic, and SMO classifiers to select the best classifier amongst them for the problem. Multi-objective PSO is used to find the optimum point from the random values of ROC plot to achieve a better quality of results. Also, C. Nimmy and K. A. Dhanya [70] propose a multi-objective PSO to improve the detection rate of RF and information gain for feature selection.

Table 11.3 summarizes the approaches used for each technique, the algorithms where the results are compared, and the evaluation measures used in comparing each technique. As observed from the table and discussed here, different ways are applied when combining RF with PSO. It is also observed that the techniques are compared with different algorithms using different evaluation measures.

11.6.3 Neural Network (NN)

A neural network simulates the brain of the humans. It is composed of nodes which reflect the biotic neurons of brain, and links to connect between the nodes.

NN accepts the data as input vector, generates the predicted results, and compares the predicted results with the labels [3].

PSO is mainly used with NN for intrusion detection to optimize the NN parameters. Other ways include feature selection, optimizing the number of hidden layer nodes, optimizing hyperboxes, and agent detection.

Optimizing NN parameters by PSO can be found in the literature in many works. RF is used for feature selection and PSO for optimizing NN parameters in the works of [33,35], as mentioned in Section 11.6.2, "Random Forest (RF)." The approach proposed in [37] considers a wavelet neural network (WNN) as the object to be trained by quantum PSO (QPSO) for mass multimedia data transmission networks. PSO is used to optimize the NN parameters in the approach proposed in [59] to identify intrusion attacks from normal ones. k-NN is then used to classify intrusion attacks into different types. Moreover, the internal power parameters, including the weight and basis, is optimized by the extreme learning machine (ELM) [4]. ELM is considered a single-layer feed-forward neural network. This work [38] introduced LE-MPSO-BP, which considers Laplacian Eigenmaps (LE) for feature selection and modified PSO to optimize weight/threshold of back propagation NN. PSO-FLN model is proposed in which PSO optimizes the weights of fast learning networks (FLN) [39]. A hybrid artificial bee colony (ABC) and PSO algorithm is used to optimize the biases and weights of NN [63]. A particle deep framework (PDF) is proposed by [77], where PSO is used to select the best values of the parameters of the deep neural network for Bot-IoT and UNSWNB15 datasets. M. Rabbani et al. (2020) [78] apply improved PSO with a back propagation neural network (IPSO-BPNN) on a UNSW-NB15 dataset to detect malicious behaviors. X. Lu et al. (2020) [54] applies PSO on a probabilistic neural network (PSO-PNN) to detect intrusion on NSL-KDD and UNSW-NB15 datasets. A feed-forward neural network with locust swarm optimization (FNNLSO) is proposed by I. Benmessahel et al. (2019) [55] to improve the intrusion detection system for NSL-KDD and UNSW-NB15 datasets.

Opposition–based PSO is also used for feature selection in the approach proposed in [3], to reduce the number of features of the NSL-KDD dataset from 41 to 22, and a probabilistic neural network (PNN) is used for intrusion detection. Other works consider PSO both for optimizing NN parameters and feature selection [5]. They propose three techniques, combining PSO with deep neural networks (DNN), long short-term memory recurrent neural networks (LSTM-RNN), and deep belief networks (DBN). These techniques are named Double PSO+DNN, Double PSO+LSTM-RNN, Double PSO+DBN, respectively. PSO is used for feature selection and hyperparameter optimization.

PSO is used with NN in many other works. The approach proposed in [56] considers the deep belief network (DBN) structure to be optimized by PSO by finding the optimal number of hidden layer nodes. A fuzzy min max NN combined with PSO for optimizing the min and max values of the hyperboxes is proposed in [65]. The approach proposed in [67] performs the election of detection agents using PSO and back propagation to detect attacks.

Table 11.4 summarizes various techniques using PSO and the purposes of each technique, and compares the evaluation measures. As observed from the table and

Privacy Preservation Tools and Techniques

TABLE 11.4

NN-based PSO techniques

Ref.	Technique	Purpose	Compared with**	Measures
[3]	OPSO- PNN	FS	PSO- NN, PSO–RB, PSO–PNN, OPSO-NN, and OPSO-RB	ACC, TPR, FPR, FNR, TNR
[4]	PSO-ELM	Optimize NN parameters	ELM	ACC, TPR, FPR, PREC, F1
[33]	PSO+NN+RF	Optimize NN parameters	k-NN, SVM, LR, DT, NB, PSO-discretize-HNB [79], MBGWO [80], MI-BGSA [81], DO-IDS [82]	ACC, PREC, FPR, FNR, TNR, F1, TPR,
[35]	PSO+NN+RF	Optimize NN parameters	CFS and SVM	TPR, FPR
[37]	QPSO-WNN	Optimize NN parameters	PSO-WNN, GA-WNN, k-NN	Time, ACC
[38]	LE-MPSO-BP	Optimize NN parameters	MPSO-BP	TPR, Time
[39]	PSO-FLN	Optimize NN parameters	ELM, sequential ELM, ATLBO-ELM, GA-ELM, PSO-ELM, HSO-ELM, FLN, GA-FLN, ATLBO-FLN	ACC
[56]	DBN-PSO	Number of hidden layer nodes	SVM, ANN, DNN, Adaboost, SA, BO	ACC, TPR, PREC, FPR
[59]	Elman NN+PSO+knn	Optimize NN parameters	Elman, ARIMA, BP	ACC, FPR
[63]	ABC+PSO+NN	Optimize NN parameters	radial basis function, voted perceptron, logistic regression, and multilayer perceptron	kappa statistic, mean absolute error, RMSE, relative absolute error, and root relative squared error
[65]	FMM PSO	Optimizing the min and max values of the hyperboxes	FMM+NN, FMM+GA, MLP, RBF, RBFN, SMO, NB, Lib-SVM, KDD, Cup Winner, KDD Cup Runner UP, FMM and FMM GA based systems	ACC, ER
[67]	PSO+NN	Agents detection	MASID	FPR, FNR, Time
[77]	PDF	Optimize NN parameters	SVM, RNN, LSTM, ARM, DT, NB, NN	ACC, PREC, TPR, FPR, FNR, F1
[78]	PSO-PNN	Optimize NN parameters	CVT, TANN, AIS, FSVM, Geometric Area Analysis, Multivariate Correlation Analysis	PREC, F1, TPR, FPR
[54]	IPSO-BPNN	Optimize NN parameters	BPNN, PSO-BPNN	ACC, FPR
[55]	FNN-LSO	Optimize NN parameters	FNN-GA, FNN-PSO	ACC, TPR, FPR, P REC, F1

** SA: simulated annealing, BO: Bayesian optimization, NB: naive Bayes, MASID: multi-agent system for intrusion detection

discussed here, different ways are applied when combining NN with PSO. Most techniques use PSO to optimize the NN parameters, while a few others consider it for different purposes such as FS, optimizing the number of hidden layer nodes, optimizing the hyperboxes, and agent detection. It is also observed that the techniques are compared with different algorithms using different evaluation measures.

11.6.4 k-Nearest Neighbor (k-NN)

Presented in the 1950s, k-NN classifies testing instances based on the closest training instances in space [90]. That is, the labels of the k closest training instances of a testing instance are identified, and the majority of their labels is considered for the testing instance as the predicted label [41]. k-NN can be defined as follows [91]:

Definition 1 *Given a set of N points P = $\{p_1, p_2...p_N\}$ in space, the $k - NN(p_i) = \{nn_1, nn_2...nn_k\}$ represents the k-nearest neighbors set of a certain point p_i, where $k = |k - NN(p_i)|$ and $k < N$, a nearest neighbor $nn_j \in \{nn_1, nn_2...nn_k\}$.*

The euclidean distance is often used to find the k closest training instances of a testing instance. The euclidean distance between point p_1 and p_2 can be defined by Equation 11.12 [92]:

$$Dist\left(p_1, p_2\right) = \sqrt{\sum_{i=1}^{F}\left(p_{1i} - p_{2i}\right)^2} \tag{11.12}$$

where F, p_{1i}, and p_{2i} are the number of features, the feature i of the first point, and the feature i of the second point, respectively.

Different uses of PSO are found in the literature: feature selection [40], optimizing NN parameters [59], and optimizing k paramesr of k-NN [41]. k-NN is then applied to predict attacks. The proposed approach in [40] uses binary PSO to select features and k-NN for classifying instances into normal or attack labels. PSO is used in [59] to optimize the parameters of Elman NN, which predicts network traffic, and k-NN is used for intrusion classification. As discussed in Section 11.6.1, PSO is used in the approach proposed in [41] to optimize the SVM RBF parameter γ and the k-NN k parameter. All these works are based on accuracy for evaluating the performance of their proposed techniques.

11.6.5 Other Related Techniques Available

There are many other techniques in the literature for intrusion detection considering PSO and other algorithms for their approach. These algorithms include MCLP [42,43,75], J48, naive Bayes and ID3 [46], SMO [34,44], principal component analysis (PCA) [71,72], Xgboost [45,58], decision tree (DT) [62,64], C4.5 and multi-class classification (MCC) [42], naive Bayes [34,46,74], IPTBK [57], k-means [69], and correlation information entropy [47]. Some others consider

Privacy Preservation Tools and Techniques

other optimization algorithms along with PSO, including Cuckoo search [60], Firefly [61], and GA [66].

11.6.6 DISCUSSION AND RECOMMENDATIONS

Since PSO is an optimizing technique, it is not used for intrusion detection as a stand-alone technique, but rather it is combined with other classification techniques to optimize their work. In intrusion detection, and as observed from the aforementioned sections, PSO is mainly used to optimize the parameters of the classification technique and is used for feature selection. PSO is mainly used for parameter selection of SVM and NN as they are used for classification. When combined with RF, PSO is mainly used for feature selection and RF for classification, but when combined with NN, RF becomes responsible for feature selection and PSO for optimizing NN parameters. Further, PSO is used with k-NN for feature selection and parameter optimization.

Most approaches either consider feature selection or parameter optimization at their proposal. Thus, it is recommended to find more works that experiment with PSO with both purposes at the same approach.

The approaches also limit their experiments with the KDD99 dataset, and its improved version NSL-KDD, while a few others consider a more recent UNSWNB15 dataset. The works by H. Kang et al. (2019) [93] and I. Ullah and Q. H. Mahmoud (2020) [94] should be considered for the recent UNSW-NB15 dataset, in addition to other recent datasets, for intrusion detection for IoT networks. It is recommended to consider different and more recent datasets, since different ways of attacks are emerging over time. In addition, recent works should not consider KDD99, which has many problems compared to the improved NSL-KDD, but rather consider the improved version.

Careful consideration should be taken of the nature of the distribution of the classes for KDD99 and NSL-KDD when evaluating the proposed approach. Both datasets are imbalanced, having approximately 20% and 80% of normal and intrusion instances, respectively. On the other hand, it is well known that the accuracy and precision measures, which are considered for evaluating many approaches, are sensitive to imbalanced data, and thus might not be reliable to measure the performance of the proposed approach [95]. The reason is that the accuracy measure could change according to the change of data distributions even if no enhancements or diminutions of the classifier are made [96].

In contrast, *TPR* and *FPR* measures are also used to evaluate the proposed techniques. *TPR* indicates how well the approach is detecting the right attacks. *FPR* is also very important, since we need to minimize the false positives from normal attacks. Both measures are suitable for imbalanced datasets and give a good indication of the quality of the proposed approach. Further measures, such as positive likelihood ($LR+$), negative likelihood ($LR-$), and geometric mean (GM), are recommended to be used to combine both *TPR* and FPR measures to give a combined evaluation of both measures for imbalanced data [96]. $LR+$ measures how much the odds of the intrusion increases when predicting intrusion instances, while $LR-$ measures how much the odds of the intrusion decrease when predicting normal instances. GM aggregates both sensitivity and specificity, so that approaches are measured with no bias of enhancing *TPR*, regardless of causing a decrease in *FPR*. *LR+*, *LR−*, and *GM* are calculated by Equations 11.13, 11.14, and 11.15 [97, 98].

$$LR+ = \frac{TPR}{FPR} \tag{11.13}$$

$$LR = \frac{FNR}{TNR} \tag{11.14}$$

$$GM = \sqrt{TPR \times FPR} \tag{11.15}$$

11.7 CONCLUSION AND FUTURE WORK

In this chapter, PSO-based techniques, common datasets, and evaluation measures for intrusion detection are discussed. PSO has been mainly used in the intrusion detection application to optimize the parameters of classification algorithms and for feature selection. NN, SVM, RF, and k-NN are widely used for PSO-based classification in this area. KDD99 and NSL-KDD are the dominating datasets used for intrusion detection with PSO. Moreover, ACC, TPR, and FPR are found to be the most used measures. Finally, further directions and insights are introduced and recommended.

For future work, experimental investigations should be considered for different ways to optimize classification algorithms in intrusion detection. New ways should be introduced and examined. Different and more recent datasets, such as the UNSW-NB15 dataset [52] and the datasets by H. Kang H. et al. (2019) [93] and I. Ullah and Q. H. Mahmoud (2020) [94] should be discovered and experimented with, on the same problem.

REFERENCES

1. N. K. Jain, U. Nangia, and J. Jain. A review of particle swarm optimization. *Journal of the Institution of Engineers (India): Series B*, 99(4): 407–411, 2018.
2. R. Qaddoura, H. Faris, I. Aljarah, and P. A. Castillo. *Evocluster: An open-source nature-inspired optimization clustering framework in Python*. In *International Conference on the Applications of Evolutionary Computation (Part of EvoStar)*, pages 20–36. Seville, Spain: Springer, 2020.
3. T. Sree Kala, and A. Christy. *An intrusion detection system using opposition based particle swarm optimization algorithm and pnn*. In *2019 International Conference on Machine Learning, Big Data, Cloud and Parallel Computing (COMITCon)*, pages 184–188. Faridabad, India: IEEE, 2019.
4. M. H. Ali, M. Fadlizolkipi, A. Firdaus, and N. Z. Khidzir. *A hybrid particle swarm optimization-extreme learning machine approach for intrusion detection system*. In *2018 IEEE Student Conference on Research and Development (SCOReD)*, pages 1–4. Selangor, Malaysia: IEEE, 2018.
5. W. Elmasry, A. Akbulut, and A. H. Zaim. Evolving deep learning architectures for network intrusion detection using a double pso metaheuristic. *Computer Networks*, 168: 107042, 2020.

6. D. J. Kalita, V. P. Singh, and V. Kumar. *Svm hyperparameters optimization using multi-pso for intrusion detection*. In *Social Networking and Computational Intelligence*, pages 227–241. Singapore: Springer, 2020.
7. N. N. El-Emam. New data-hiding algorithm based on adaptive neural networks with modified particle swarm optimization. *Computers & Security*, 55: 21–45, 2015.
8. M. Habib, I. Aljarah, H. Faris, and S. Mirjalili. *Multiobjective particle swarm optimization for botnet detection in internet of things*. In S. Mirjalilli, H. Faris, and I. Aljarah (Eds.), *Evolutionary Machine Learning Techniques*, pages 203–229. Singapore: Springer, 2020.
9. H. Faris, I. Aljarah, and B. Al-Shboul. *A hybrid approach based on particle swarm optimization and random forests for e-mail spam filtering*. In *International Conference on Computational Collective Intelligence*, pages 498–508. Da Nang, Vietnam: Springer, 2016.
10. M. Habib, I. Aljarah, H. Faris, and S. Mirjalili. *Multiobjective particle swarm optimization: Theory, literature review, and application in feature selection for medical diagnosis*. In S. Mirjalili, H. Faris, and I. Aljarah (Eds.), *Evolutionary Machine Learning Techniques*, pages 175–201. Singapore: Springer, 2020.
11. Y. Hernafi, M. B. Ahmed, and M. Bouhorma. Aco and pso algorithms for developing a new communication model for vanet applications in smart cities. *Wireless Personal Communications*, 96(2): 2039–2075, 2017.
12. Y. Zhang, S. Wang, and G. Ji. A comprehensive survey on particle swarm optimization algorithm and its applications. In Y. Zhang, S. Balochian, P. Agarwal, V. Bhatnager, and O. J. Houshey (Eds.), *Mathematical Problems in Engineering*, pages 1–38. 2015.
13. M. Eslami, H. Shareef, M. Khajehzadeh, and A. Mohamed. A survey of the state of the art in particle swarm optimization. *Research Journal of Applied Sciences, Engineering and Technology*, 4(9): 1181–1197, 2012.
14. S. Sengupta, S. Basak, and R. A. Peters. Particle swarm optimization: A survey of historical and recent developments with hybridization perspectives. *Machine Learning and Knowledge Extraction*, 1(1): 157–191, 2019.
15. Y. Liu, X. Ling, Z. Shi, M. Lv, J. Fang, and Z. Liang. A survey on particle swarm optimization algorithms for multimodal function optimization. *JSW*, 6(12): 2449–2455, 2011.
16. G. Armano, and M. R. Farmani. Multiobjective clustering analysis using particle swarm optimization. *Expert Systems with Applications*, 55: 184–193, 2016.
17. S. Lalwani, H. Sharma, S. C. Satapathy, K. Deep, and J. C. Bansal. A survey on parallel particle swarm optimization algorithms. *Arabian Journal for Science and Engineering*, 44(4): 2899–2923, 2019.
18. S. Sarkar, A. Roy, and B. S. Purkayastha. Application of particle swarm optimization in data clustering: A survey. *International Journal of Computer Applications*, 65(25): 38–46. 2013.
19. X. Li, S. Qing, H. Wen, C. Liu, and S. Xu. An overview of cluster analysis based on particle swarm optimization. *Journal of Computational and Theoretical Nanoscience*, 13(11): 8604–8614, 2016.
20. S. Rana, S. Jasola, and R. Kumar. A review on particle swarm optimization algorithms and their applications to data clustering. *Artificial Intelligence Review*, 35(3): 211–222, 2011.
21. V. Mangat. *Survey on particle swarm optimization based clustering analysis*. In *Swarm and Evolutionary Computation*, pages 301–309. Zakopane, Poland: Springer, 2012.
22. A. A. A. Esmin, R. A. Coelho, and S. Matwin. A review on particle swarm optimization algorithm and its variants to clustering high-dimensional data. *Artificial Intelligence Review*, 44(1): 23–45, 2015.

23. P. Vora, B. Oza, et al. A survey on k-mean clustering and particle swarm optimization. *International Journal of Science and Modern Engineering*, 1(3): 24–26, 2013.

24. S. L. Marie-Sainte. A survey of particle swarm optimization techniques for solving university examination timetabling problem. *Artificial Intelligence Review*, 44(4): 537–546, 2015.

25. R. A. Jamous, E. Seidy, A. A. Tharwat, and Bayoum Ibrahim Bayoum. Modifications of particle swarm optimization techniques and its application on stock market: A survey. *International Journal of Advanced Computer Science and Applications (IJACSA)*, 6(3): 99–108, 2015.

26. R. S. M. Lakshmi Patibandla, B. T. Rao, P. S. Krishna, and V. R. Maddumala. *Medical data clustering using particle swarm optimization method. Journal of Critical Reviews*, 7(6): 363–367, 2020.

27. Q. Pi, and H. Ye. *Survey of particle swarm optimization algorithm and its applications in antenna circuit*. In *2015 IEEE international conference on communication problem-solving (ICCP)*, pages 492–495. Guilin, China: IEEE, 2015.

28. M. Elbes, S. Alzubi, T. Kanan, A. Al-Fuqaha, and B. Hawashin. A survey on particle swarm optimization with emphasis on engineering and network applications. *Evolutionary Intelligence*, 12: 113–129, 2019.

29. R. V. Kulkarni, and G. K. Venayagamoorthy. Particle swarm optimization in wireless-sensor networks: A brief survey. *IEEE Transactions on Systems, Man, and Cybernetics, Part C (Applications and Reviews)*, 41(2): 262–267, 2010.

30. Y. Bi. A hybrid pso-svm model for network intrusion detection. *International Journal of Security and Networks*, 11(4): 196–203, 2016.

31. B. Tan, Y. Tan, and Y. X. Li. Research on intrusion detection system based on improved pso-svm algorithm. *Chemical Engineering Transactions*, 51: 583–588, 2016.

32. N. Kunhare, R. Tiwari, and J. Dhar. Particle swarm optimization and feature selection for intrusion detection system. *Sadhana*, 45(109): 1–14, 2020.

33. A. Dickson, and C. Thomas. Improved pso for optimizing the performance of intrusion detection systems. *Journal of Intelligent & Fuzzy Systems*, 38(Preprint): 1–11, 2020.

34. H. Li, W. Guo, G. Wu, and Y. Li. *A rf-pso based hybrid feature selection model in intrusion detection system*. In *2018 IEEE Third International Conference on Data Science in Cyberspace (DSC)*, pages 795–802. Guangzhou, China: IEEE, 2018.

35. B. A. Tama, and K.-H. Rhee. *An integration of pso-based feature selection and random forest for anomaly detection in iot network*. In *MATEC Web of Conferences*, volume 159, page 01053. Indonesia: EDP Sciences, 2018.

36. L. Guo. Research on anomaly detection in massive multimedia data transmission network based on improved pso algorithm. *IEEE Access*, 8: 95368–95377, 2020.

37. Y. Liu, D. Qiu, and H. Li. *The intrusion detection modle utilizing le and modified pso-bp*. In *2017 8th IEEE International Conference on Software Engineering and Service Science (ICSESS)*, pages 318–321. Beijing, China: IEEE, 2017.

38. M. H. Ali, B. A. D. Al Mohammed, A. Ismail, and M. F. Zolkipli. A new intrusion detection system based on fast learning network and particle swarm optimization. *IEEE Access*, 6: 20255–20261, 2018.

39. A. R. Syarif, and W. Gata. Intrusion detection system using hybrid binary pso and k-nearest neighborhood algorithm. In *2017 11th International Conference on Information & Communication Technology and System (ICTS)*, pages 181–186. Surabaya, Indonesia: IEEE, 2017.

40. A. A. Aburomman, and M. B. I. Reaz. A novel svm-knn-pso ensemble method for intrusion detection system. *Applied Soft Computing*, 38: 360– 372, 2016.

Privacy Preservation Tools and Techniques

41. A. M. V. Bharathy, and A. M. Basha. A multi-class classification mclp model with particle swarm optimization for network intrusion detection. *Sadhana*, 42(5): 631–640, 2017.

42. S. M. H. Bamakan, H. Wang, T. Yingjie, and Y. Shi. An effective intrusion detection framework based on mclp/svm optimized by timevarying chaos particle swarm optimization. *Neurocomputing*, 199: 90–102, 2016.

43. M. Moukhafi, K. El Yassini, and S. Bri. A novel anomaly intrusion detection based on smo optimized by pso with pre-processing of data set. *J. Mobile Multimedia*, 13(3&4): 197–209, 2017.

44. H. Jiang, Z. He, G. Ye, and H. Zhang. Network intrusion detection based on pso-xgboost model. *IEEE Access*, 8: 58392–58401, 2020.

45. J. Wang, and Y. Jin. *A novel feature-selection approach based on particle swarm optimization algorithm for intrusion detection systems (workshop paper)*. In *International Conference on Collaborative Computing: Networking, Applications and Worksharing*, pages 455–465. London, UK: Springer, 2019.

46. Y.-F. Wang, P.-Y. Liu, M. Ren, and X.-X. Chen. *Intrusion detection algorithms based on correlation information entropy and binary particle swarm optimization*. In *2017 13th International Conference on Natural Computation, Fuzzy Systems and Knowledge Discovery (ICNC-FSKD)*, pages 2829–2834. Guilin, China: IEEE, 2017.

47. J. Kennedy, and R. Eberhart. *Particle swarm optimization*. In *Proceedings of ICNN'95-International Conference on Neural Networks*, volume 4, pages 1942–1948. Perth, WA, Australia: IEEE, 1995.

48. R. Qaddoura, W. Al Manaseer, M. A. M. Abushariah, and M. A. Alshraideh. Dental radiography segmentation using expectationmaximization clustering and grasshopper optimizer. *Multimedia Tools and Applications*, 79: 22027–22045, 2020.

49. S.-W. Lin, K.-C. Ying, C.-Y. Lee, and Z.-J. Lee. An intelligent algorithm with feature selection and decision rules applied to anomaly intrusion detection. *Applied Soft Computing*, 12(10): 3285–3290, 2012.

50. M. Tavallaee, E. Bagheri, W. Lu, and A. A. Ghorbani. *A detailed analysis of the kdd cup 99 data set*. In *2009 IEEE symposium on computational intelligence for security and defense applications*, pages 1–6. Barcelona, Spain: IEEE, 2009.

51. N. Moustafa, and J. Slay. *Unsw-nb15: a comprehensive data set for network intrusion detection systems (unsw-nb15 network data set)*. In *2015 military communications and information systems conference (MilCIS)*, pages 1–6. Canberra, Australia: IEEE, 2015.

52. B. A. Tama, M. Comuzzi, and K.-H. Rhee. Tse-ids: A two-stage classifier ensemble for intelligent anomaly-based intrusion detection system. *IEEE Access*, 7: 94497–94507, 2019.

53. X. Lu, D. Han, L. Duan, and Q. Tian. Intrusion detection of wireless sensor networks based on ipso algorithm and bp neural network. *International Journal of Computational Science and Engineering*, 22(2–3): 221–232, 2020.

54. I. Benmessahel, K. Xie, M. Chellal, and T. Semong. A new evolutionary neural networks based on intrusion detection systems using locust swarm optimization. *Evolutionary Intelligence*, 12(2): 131–146, 2019.

55. X. Tan, S. Su, Z. Zuo, X. Guo, and X. Sun. Intrusion detection of uavs based on the deep belief network optimized by pso. *Sensors*, 19(24): 5529, 2019.

56. H.-C. Lin, P. Wang, and W.-H. Lin. Implementation of a pso-based security defense mechanism for tracing the sources of ddos attacks. *Computers*, 8(4): 88, 2019.

57. J. Wang, C. Liu, X. Shu, H. Jiang, X. Yu, J. Wang, and W. Wang. *Network intrusion detection based on xgboost model improved by quantumbehaved particle swarm optimization*. In *2019 IEEE Sustainable Power and Energy Conference (iSPEC)*, pages 1879–1884. Beijing, China: IEEE, 2019.

58. W. Guoli. *Traffic prediction and attack detection approach based on pso optimized elman neural network*. In *2019 11th International Conference on Measuring Technology and Mechatronics Automation (ICMTMA)*, pages 504–508. Qiqihar, China: IEEE, 2019.
59. P. Ghosh, A. Karmakar, J. Sharma, and S. Phadikar. *Cs-pso based intrusion detection system in cloud environment*. In A. Abraham, P. Dutta, J. K. Mandal, A. Bhattacharya, and S. Dutta (Eds.), *Emerging Technologies in Data Mining and Information Security*, pages 261–269. Singapore: Springer, 2019.
60. R. Kondaiah, and B. Sathyanarayana. Trust factor and fuzzy-firefly integrated particle swarm optimization based intrusion detection and prevention system for secure routing of manet. *International Journal of Computer Sciences and Engineering*, 10(1): 13–33, 2018.
61. L. Li, Y. Yu, S. Bai, J. Cheng, and X. Chen. Towards effective network intrusion detection: A hybrid model integrating gini index and gbdt with pso. *Journal of Sensors*, 2018: 1–9, 2018.
62. W. A. H. M. Ghanem, and A. Jantan. New approach to improve anomaly detection using a neural network optimized by hybrid abc and pso algorithms. *Pakistan Journal of Statistics*, 34(1): 1–14, 2018.
63. A. J. Malik, and F. A. Khan. A hybrid technique using binary particle swarm optimization and decision tree pruning for network intrusion detection. *Cluster Computing*, 21(1): 667–680, 2018.
64. C. Azad, and V. K. Jha. Fuzzy min–max neural network and particle swarm optimization based intrusion detection system. *Microsystem Technologies*, 23(4): 907–918, 2017.
65. K. Anusha and E. Sathiyamoorthy. A decision tree-based rule formation with combined pso-ga algorithm for intrusion detection system. *International Journal of Internet Technology and Secured Transactions*, 6(3): 186–202, 2016.
66. R. K. Cherian, and A. S. Nargunam. Distributed agent-based detection system using pso and neural network for manet. *International Journal of Mobile Network Design and Innovation*, 6(4): 185–195, 2016.
67. A. J. Malik, W. Shahzad, and F. A. Khan. Network intrusion detection using hybrid binary pso and random forests algorithm. *Security and Communication Networks*, 8(16): 2646–2660, 2015.
68. S.-H. Li, Y.-C. Kao, Z.-C. Zhang, Y.-P. Chuang, and D. C. Yen. A network behavior-based botnet detection mechanism using pso and k-means. *ACM Transactions on Management Information Systems (TMIS)*, 6(1): 1–30, 2015.
69. N. Cleetus, and K. A. Dhanya. Multi-objective particle swarm optimization in intrusion detection. In L. C. Jain, H. S. Behera, J. K. Mandal, and D. P. Mohapatra (Eds.), *Computational Intelligence in Data Mining-Volume 2*, pages 175–185. New Delhi: Springer, 2015.
70. I. Ahmad. Feature selection using particle swarm optimization in intrusion detection. *International Journal of Distributed Sensor Networks*, 11(10): 806954, 2015.
71. F. Kuang, S. Zhang, Z. Jin, and W. Xu. A novel svm by combining kernel principal component analysis and improved chaotic particle swarm optimization for intrusion detection. *Soft Computing*, 19(5): 1187–1199, 2015.
72. G. P. Wang, S. Y. Chen, and J. Liu. Anomaly-based intrusion detection using multiclass-svm with parameters optimized by pso. *International Journal of security and its Applications*, 9(6): 227–242, 2015.
73. A. Dickson, and C. Thomas. *Optimizing false alerts using multi-objective particle swarm optimization method*. In *2015 IEEE International Conference on Signal Processing, Informatics, Communication and Energy Systems (SPICES)*, pages 1–5. Kerala, India: IEEE, 2015.

74. S. M. H. Bamakan, B. Amiri, M. Mirzabagheri, and Y. Shi. A new intrusion detection approach using pso based multiple criteria linear programming. *Procedia Computer Science*, 55: 231–237, 2015.

75. B. A. Tama, and K. H. Rhee. *A combination of pso-based feature selection and tree-based classifiers ensemble for intrusion detection systems.* In D. S. Park, H. C. Chao, Y. S. Jeong, and J. J. Park (Eds.), *Advances in Computer Science and Ubiquitous Computing*, pages 489–495. Springer, 2015.

76. N. Koroniotis, N. Moustafa, and E. Sitnikova. A new network forensic framework based on deep learning for internet of things networks: A particle deep framework. *Future Generation Computer Systems*, 110: 91–106, 2020.

77. M. Rabbani, Y. L. Wang, R. Khoshkangini, H. Jelodar, R. Zhao, and P. Hu. A hybrid machine learning approach for malicious behaviour detection and recognition in cloud computing. *Journal of Network and Computer Applications*, 151: 102507, 2020.

78. A. Elngar, D. A. E. A. Mohamed, and F. Ghaleb. A real-time anomaly network intrusion detection system with high accuracy. *Information Sciences Letters*, 2(2): 49–56, 2013.

79. Q. M. Alzubi, M. Anbar, Z. N. M. Alqattan, M. A. AlBetar, and R. Abdullah. Intrusion detection system based on a modified binary grey wolf optimisation. *Neural Computing and Applications*, 32: 6125–6137, 2020.

80. H. Bostani, and M. Sheikhan. Hybrid of binary gravitational search algorithm and mutual information for feature selection in intrusion detection systems. *Soft Computing*, 21(9): 2307–2324, 2017.

81. J. Ren, J. Guo, Q. Wang, H. Yuan, X. Hao, and J. Hu. Building an effective intrusion detection system by using hybrid data optimization based on machine learning algorithms. *Security and Communication Networks*, 2019: 1–11, 2019.

82. S. Chebrolu, A. Abraham, and J. P. Thomas. Feature deduction and ensemble design of intrusion detection systems. *Computers & Security*, 24(4): 295–307, 2005.

83. N. Shone, T. N. Ngoc, V. D. Phai, and Q. Shi. A deep learning approach to network intrusion detection. *IEEE Transactions on Emerging Topics in Computational Intelligence*, 2(1): 41–50, 2018.

84. S. Potluri, and C. Diedrich. *Accelerated deep neural networks for enhanced intrusion detection system.* In *2016 IEEE 21st international conference on emerging technologies and factory automation (ETFA)*, pages 1–8. Berlin, Germany: IEEE, 2016.

85. C. Yin, Y. Zhu, J. Fei, and X. He. A deep learning approach for intrusion detection using recurrent neural networks. *IEEE Access*, 5: 21954–21961, 2017.

86. M. Z. Alom, V. R. Bontupalli, and T. M. Taha. *Intrusion detection using deep belief networks.* In *2015 National Aerospace and Electronics Conference (NAECON)*, pages 339–344. Dayton, OH: IEEE, 2015.

87. L. Yang, Y. Q. Zhang, X. R. Jin, S. Zhang, and Y. P. Lee. Dualband infrared perfect absorber for plasmonic sensor based on the electromagnetically induced reflection-like effect. *Optics Communications*, 371: 173–177, 2016.

88. J. Han, J. Pei, and M. Kamber. *Data mining: Concepts and techniques.* Waltham, MA: Elsevier, 2011.

89. R. Qaddoura, H. Faris, and I. Aljarah. An efficient clustering algorithm based on the k-nearest neighbors with an indexing ratio. *International Journal of Machine Learning and Cybernetics*, 11(3): 675–714, 2020.

90. R. Qaddoura, H. Faris, and I. Aljarah. An efficient evolutionary algorithm with a nearest neighbor search technique for clustering analysis. *Journal of Ambient Intelligence and Humanized Computing*, 2020. DOI: https://doi.org/10.1007/s12652-020-02570-2

91. H. Kang, D. H. Ahn, G. M. Lee, J. D. Yoo, K. Ho Park; H. K. Kim. *IoT network intrusion dataset*, 2019. http://dx.doi.org/10.21227/q70p-q449. Accessed Oct 8 2020.

92. I. Ullah, and Q. H. Mahmoud. *A scheme for generating a dataset for anomalous activity detection in iot networks*. In *Canadian Conference on Artificial Intelligence*, pages 508–520. Ottowa, ON, Canada: Springer, 2020.
93. T. Fawcett. An introduction to roc analysis. *Pattern Recognition Letters*, 27(8): 861–874, 2006.
94. A. Tharwat. Classification assessment methods. *Applied Computing and Informatics*, 17(1): 168–192, 2020.
95. M. Sokolova, N. Japkowicz, and S. Szpakowicz. *Beyond accuracy, fscore and roc: a family of discriminant measures for performance evaluation*. In *Australasian Joint Conference on Artificial Intelligence*, pages 1015–1021. Hobart, TAS, Australia: Springer, 2006.
96. S. Boughorbel, F. Jarray, and M. El-Anbari. Optimal classifier for imbalanced data using matthews correlation coefficient metric. *PLoS One*, 12(6): e0177678, 2017.

12 Web Security Vulnerabilities

Identification, Exploitation, and Mitigation

Sachin Kumar Sharma
Manipal University Jaipur; Dr CBS Cybersecurity Services LLP, India

Dr. Arjun Singh, Dr. Punit Gupta and Dr. Vijay Kumar Sharma
Manipal University Jaipur, India

CONTENTS

12.1	Introduction	184
12.2	Introduction to Important Web Vulnerabilities	185
12.3	Injection	186
	12.3.1 Types of Injections and Their Working Details	186
12.4	Identification of SQL Injection Vulnerabilities	187
12.5	Mitigation of Injection Vulnerabilities	199
12.6	Broken Authentication and Session Management	199
	12.6.1 Exploitation of Broken Authentication Vulnerability	200
	12.6.2 Mitigation of Broken Authentication and Session Management	208
12.7	Sensitive Data Exposure	209
	12.7.1 Identification of Sensitive Data Exposure Vulnerability	209
	12.7.2 Mitigation of Sensitive Data Exposure Vulnerability	210
12.8	External Entities of XML (XXE)	210
	12.8.1 Identification of XML External Entities Vulnerability	210
	12.8.2 Mitigation of XML External Entities	211
12.9	Broken Access Control	211
	12.9.1 Identification of Broken Access Control Vulnerability	211
	12.9.2 Prevent and Mitigate Broken Access Control	212
12.10	Misconfiguration of Security Options	212
	12.10.1 Identification of Security Misconfiguration Vulnerabilities	212
	12.10.2 Identification and Mitigation of Cross-Site Scripting (XSS)	213

DOI: 10.1201/9781003145042-12

12.10.3	Mitigation and Prevention of XSS Vulnerability	213
12.10.4	Identification and Mitigation of Insecure Deserialization	213
12.11	Identification of Insecure Deserialization	214
12.11.1	Prevention of Insecure Deserialization	214
12.11.2	Identification and Mitigation of Using Components with Known Vulnerabilities	214
12.11.3	Identification of Components with Known Vulnerabilities	214
12.11.4	Mitigation of Components with Known Vulnerabilities	215
12.12	Insufficient Logging and Monitoring Vulnerabilities	215
12.12.1	Identification of Insufficient Logging and Monitoring Vulnerabilities	216
12.12.2	Insufficient Logging and Monitoring Vulnerabilities	216
12.13	Web Security Standards	216
12.14	Conclusion	217
References		218

12.1 INTRODUCTION

At the present time, all the data or information of most of organizations is computer resource generated, computer resource processed, computer resource transferred, and stored in the memory of the computer resource system. Computer resources such as desktops, laptops, communication devices, networking devices, etc. are used to facilitate communication activities through the Internet.

Most organizations take advantage of computer communication technology to enhance their business activities; simultaneously, they need data protection. For this, they are adopting scybersecurity precautions to prevent data protection issues. Organization protect computer resources with the best firewall, intrusion detection systems, intrusion prevention systems, end point security, server security, analytics tools etc. The biggest challenge for an organization is to prevent their web applications or portals from unauthorized activities, because web applications are available 24/7 through the Internet for all users.

Using the Internet, an organization can represent its services to their clients 24/7, which is very important for businesses these days. But on the other hand, if any organization's web application has vulnerabilities or loopholes, then it will make the organization vulnerable to hacking. Cybercriminals can hack important data and business information from it. Cybercriminals take advantage of different IT security vulnerabilities in a website/portal and exploit that vulnerability to hack the web application server. Cybercriminals can deface or morph the web applications and modify, delete, alter, or destroy personal sensitive data from the web servers. The impact of hacking in the web servers may result in the loss of reputation of the brand, the capital and revenue of the organization. This raises the question, what is vulnerability? It is a weakness, loophole, or flaw in any computer resource. If it is exploited, then it could result in unauthorized access, denial of service (DoS), or other possible web application attacks on the computer resource.

It is essential for an organization to identify the web security risks in their web applications. To minimize web application risks, the organization should able to

Web Security Vulnerabilities
185

identify the vulnerabilities in the web applications. As per the modus operandi of the cybercriminals and various web security standards, the following vulnerabilities are critical for any web application to address.

12.2 INTRODUCTION TO IMPORTANT WEB VULNERABILITIES

a. **Injection:** An injection flaw occurs when a random input or command provides an interpreter easy access to sensitive data without proper authentication and authorization. Some common popular web application injections are: structured query language (SQL) injection, operating system (OS) injection, and lightweight directory access protocol (LDAP) injection.

b. **Broken Authentication and Session Management:** If functions related to authentication and session management for any web application are implemented incorrectly, then vulnerability arises, related to broken authentication and session management. For example, the web application allows attackers to compromise passwords, keys, or session tokens or to exploit other implementation flaws to assume other users' identities temporarily or permanently.

c. **Sensitive Data Exposure:** Sometimes, web applications do not protect sensitive data and information, for example, financial, medical, health, and personal identifiable information. Attackers may hack, delete, or modify such weakly protected data to conduct financial fraud, phishing, identity theft, or other cybercrimes. Weak or missing encryption may result in sensitive data exposure.

d. **External Entities of XML:** Some non-updated misconfigured XML content evaluates external entity references within the XML documents. External entities can be used to disclose internal data or information files using the URL, internal file sharing mechanism, internal port scanning, remote code execution, and DoS attacks.

e. **Broken Access Control:** Weak and missing controls on what authenticated users are allowed/not allowed to do are often misconfigured.

f. **Misconfiguration of Security Options:** This is the most commonly seen issue in web applications and portals. Using the insecure or by default configurations, incomplete configurations may result in security breach.

g. **Insecure Deserialization:** This is a tampering flaw, which often results in the remote execution of various codes, scripts, etc. This flaw can be used by cybercriminals to perform various web attacks, such as replay, privilege escalation, injection attacks, etc.

h. **Using Interfaces (Components) with Known Vulnerabilities:** Web applications use the components such as libraries, frameworks, software modules, open source codes, scripts, etc.

i. **Insufficient Logging and Monitoring:** This flaw occurs when a web application does not have proper logging and monitoring functions. If unauthorized activity occurs on the web application, it will trace this activity. It allows attackers to further attack the web application and tamper with, extract, or destroy sensitive data and information.

186   Cybersecurity

We now cover in detail each vulnerability, its type, its exploitability, and its prevention.

12.3   INJECTION

A web application may be vulnerable to an injection attack if:

a) User-supplied different input is not properly validated, sanitized, or filtered.
b) Random dynamic commands and queries are used directly as input to the interpreter.
c) Random queries or hostile data is used in the textbox or search field to extract additional sensitive data or information.
d) Hostile data is concatenated or used directly as different random SQL queries or commands, which have the structure and hostile data in commands, dynamic queries, or stored procedures.

12.3.1   TYPES OF INJECTIONS AND THEIR WORKING DETAILS

Injections can be classified as per the type of random supplied input in them. Some common injections are:

a) SQL injection
b) NoSQL injection
c) OS command injection
d) Object graph navigation library (OGNL) injection and expression language (EL) injection

Injection can result in data loss, the corruption or deletion of data, disclosure to unauthorized users, loss of accountability, reduced availability, etc.

a) **SQL Injection:** This flaw allows attackers to interfere with different random SQL queries to a web application database. A successful SQL injection input may exploit the database of the web application and read important sensitive data and information from the database, change the database, modify the database, etc. Hackers may use SQL commands to execute administration operations on the database. SQL injection vulnerability can be classified as:
   • **In-band SQL injection**: This occurs when an attacker is able to use the same channel of communication to perform the attack and gather the results. It can be further categorized as error-based SQL injection and union-based SQL injection.
       ○ **Error-based SQL injection**
       ○ **Union-based SQL injection**.
   • **Inferential/Blind SQL injection:** In this attack, the attacker would not be able to see the result, as in a blind SQL injection. An attacker is able to reconstruct the structure of the database by sending different payloads and observing the response of the web application and the resulting behavior of

Web Security Vulnerabilities 187

the database. The inferential SQL injection can be further categorized as following two injections:
- ○ **Blind Boolean-based SQL injection**
- ○ **Blind time-based SQL injection**
- **Out-of-band SQL injection:** This occurs when an attacker is unable to use the same channel to perform the attack and gather the results. In this injection, the attacker depends on the ability of database server to make HTTP or DNS requests, to deliver the data to an attacker.

b) **NoSQL Injection:** This refers to non-relational databases such as MongoDB, an increasingly common distributed back-end for general purposes. It is a document-based database for modern software developers and for the distributed cloud and web applications age.

c) **OS command Injection:** It is a web vulnerability that allows an attacker to execute operating system (OS) commands on the web application and can compromise the web application and can get all its available data and information. An attacker can exploit this vulnerability to compromise the hosting IT infrastructure, details of installed operating system and exploit the other system within the organization.

An OS command injection is possible when an web application allows random user supplied inputs such as forms, cookies, HTTP headers, other OS commands, etc. to a system shell. These different commands are executed with the privileges of the vulnerable web application. These attacks mainly occur due to insufficient input validation in the web applications.

12.4 IDENTIFICATION OF SQL INJECTION VULNERABILITIES

The following are some attackers' modus operandi to identify SQL injection vulnerabilities:

a) Untrusted data in an web application in the construction of the following vulnerable SQL query:

```
String query = "SELECT * FROM user WHERE ID='" +
request.getParameter("id") + "'";
```

b) Similarly, an application may result in queries that are still vulnerable, such as:

```
Query SQL = session.createQuery("FROM accounts WHERE
ID='" + request.getParameter("id") + "'");
```

c) In both cases, the attacker changes the "id" parameter value and sends: 'or' '0'='0. For example:

```
http://xyz.com/session/userView?id=' or '0'='0
```

d) Similarly, the attacker modifies the "password" parameter value and send similar input: ' or '0'='0

e) If the web application is SQL injection vulnerable, then the "or" part becomes true as or '0'='0 and bypasses the user authentication.

f) In another example that examines the effects of a different malicious value passed to the SQL query, if an attacker with the username "abc" enters the string as:

SELECT * FROM products WHERE owner = 'abc' AND productname = 'name'; DELETE FROM items;

g) An error-based SQL injection may also identified in the web application, which has dynamic values. We are using a vulnerable web application of the Acunetix tool (http://testphp.vulnweb.com/), see Figure 12.1.

FIGURE 12.1 Test vulnerable web application (http://testphp.vulnweb.com/).

h) Identify the error-based injection as shown above shown vulnerable web application, for this we navigate to dynamic value in the URL **cat=1** (listproducts.php?cat=1), Figure 12.2.

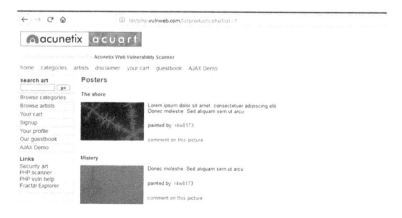

FIGURE 12.2 Identify the dynamic value in the vulnerable web application (cat = 1).

Web Security Vulnerabilities 189

i) Insert just a single quote after dynamic value **cat=1'** to check the possibility of error-based SQL injection, Figure 12.3.

FIGURE 12.3 Insert single quote (') after dynamic value.

j) Database error reflected on the browser, there is a possibility of error based SQL injection in the web application, Figure 12.4.

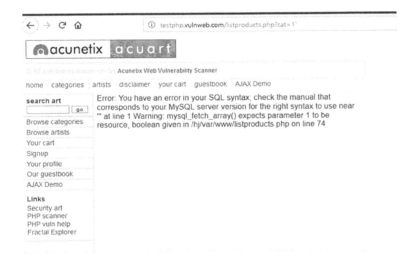

FIGURE 12.4 Database error of SQL syntax.

To identify an OS injection, we use the following commands to execute different OS commands for identification of operating-system-related details of the web server:

a) **127.0.0.1|whoami** shows you the name of the user which is currently running
b) **127.0.0.1|uname -a** shows the running web server operating the system version
c) **127.0.0.1&&ifconfig** shows all the network-configuration-related information
d) **127.0.0.1&&php -v** provides a version of PHP running on the web applications server.
e) **127.0.0.1 && cat/etc/passwd** displays the password directory of users on the backend Linux Server.
f) **127.0.0.1&&/etc/shadow** displays all the hashed passwords, if you are running with root privileges.

The following is a step-by step-procedure to identify an OS-based injection. For it, we used an open source tool of vulnerability testing bed named OWASP Mutilldae II, in the web application of the Open Web Application Security Project (OWASP):

1. First, we install OWASP Mutillidae II in a local server application and run it locally (127.0.0.1/mutilldae), Figure 12.5

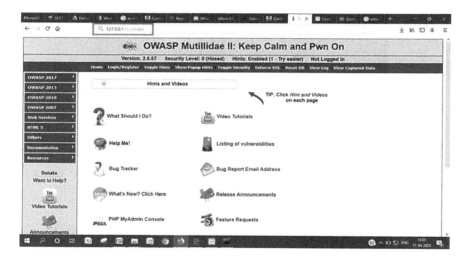

FIGURE 12.5 OWASP mutillidae II interface.

2. Go to OWASP 2017 Section: A1Injections (Others): Command Injection: DNS Lookup Section, Figure 12.6

Web Security Vulnerabilities 191

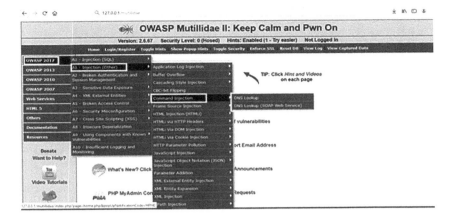

FIGURE 12.6 Command injection interface.

3. DNS lookup command injection interface, where an input box of host name/IP can be used to get important OS information, Figure 12.7

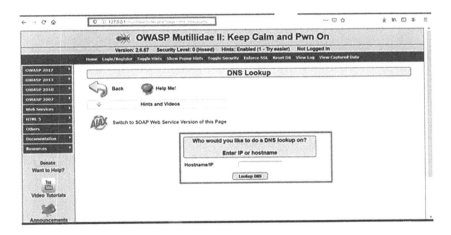

FIGURE 12.7 DNS lookup interface.

4. Check the ping command of the local host IP (127.0.0.1), Figure 12.8

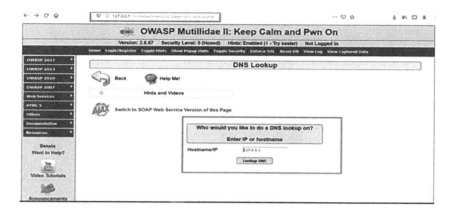

FIGURE 12.8 Lookup DNS for localhost 127.0.0.1.

5. Results of local host ping 127.0.0.1, where the WAN IP address of machine is received as follows: **2409:4052:219f:df75:642b:4ff:fee4:b9a7,** Figure 12.9

FIGURE 12.9 Result of ping command for 127.0.0.1.

Web Security Vulnerabilities 193

6. whoami OS command to check the current user using OS command injection. The command is 127.0.0.1|whoami, Figure 12.10

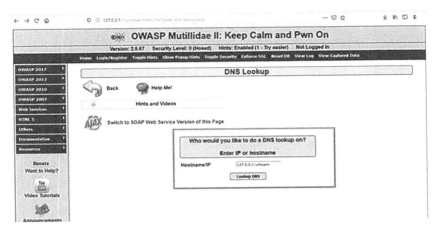

FIGURE 12.10 OS command injection 127.0.0.1 | whoami.

7. Output of OS Command injection **127.0.0.1|whoami, Host name: drcbs-lt03 and user: sachin,** Figure 12.11

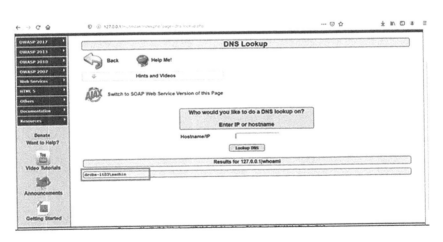

FIGURE 12.11 Host name and username using OS command injection.

8. Run the ipconfig OS command on the web application to get IP, subnet, gateway, and other IP configuration details. For this, OS command **127.0.0.1&&ipconfig is** used, Figure 12.12

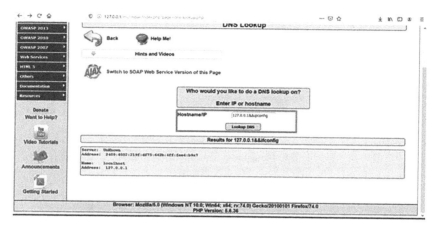

FIGURE 12.12 OS command injection 127.0.0.1&&ipconfig.

9. Output of OS Command injection **127.0.0.1&&ipconfig, with IP address, local IP address of machine, subnet mask, default gateway, etc.,** Figure 12.13

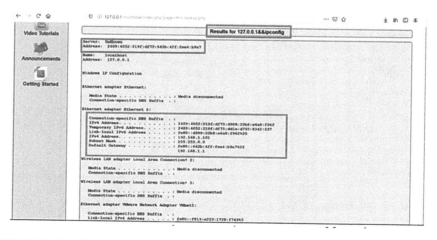

FIGURE 12.13 Output of OS command injection 127.0.0.1&&ipconfig.

Web Security Vulnerabilities   195

10. **getmac OS command to check the physical addresses (MAC Addresses) of the target machine. The command is 127.0.0.1&&getmac,** Figure 12.14

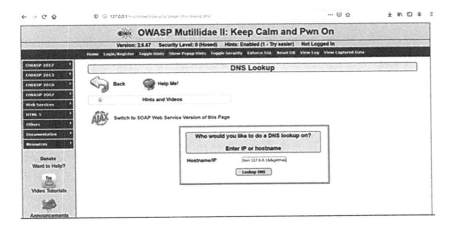

FIGURE 12.14 OS command injection 127.0.0.1&&getmac.

11. Output of OS Command injection **127.0.0.1&&getmac, in which the MAC addresses of the machine** Figure 12.15

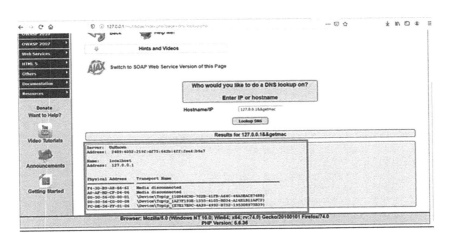

FIGURE 12.15 output of OS command injection 127.0.0.1&&getmac.

12. Command prompt access through OS command injection **127.0.0.1&&cmd,** Figure 12.16

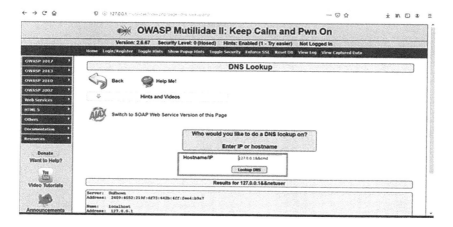

FIGURE 12.16   OS command injection 127.0.0.1&&cmd.

13. Output of OS Command injection **127.0.0.1&&cmd, in which the command prompt of the machine is reflected the on web page,** Figure 12.17

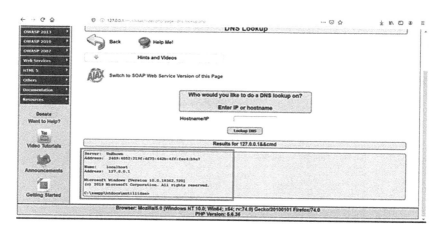

FIGURE 12.17   Output of OS command injection 127.0.0.1&&cmd.

Web Security Vulnerabilities   197

14. systeminfo OS command to get information of the target machine using **127.0.0.1&&systeminfo**, Figure 12.18

FIGURE 12.18 OS command injection 127.0.0.1&&systeminfo.

15. Output of OS Command injection **127.0.0.1&&systeminfo, with the configuration information of the machine, such as host name, operating system, OS version, system manufacturer, system model, domain, available physical memory (RAM), hotfix, etc. Using this OS command injection, an attacker can identify the configuration information of web server,** Figure 12.19

FIGURE 12.19 Output of OS command injection 127.0.0.1&&systeminfo.

16. OS command injection task list **to check the currently running tasks on the target machine. The command is 127.0.0.1&&tasklist,** Figure 12.20

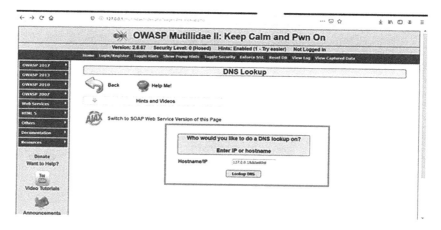

FIGURE 12.20 OS command injection to check currently running task on target machine.

17. Output of OS Command injection **127.0.0.1&&tasklist and** OS command injection **dir,** to find the whole directory of web pages located on the server, Figure 12.21

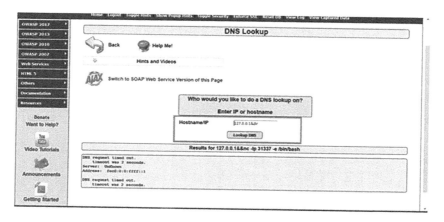

FIGURE 12.21 OS command injection to find whole directory of web pages located in web server.

Web Security Vulnerabilities

18. Output of OS Command injection **127.0.0.1&&dir,** Figure 12.22

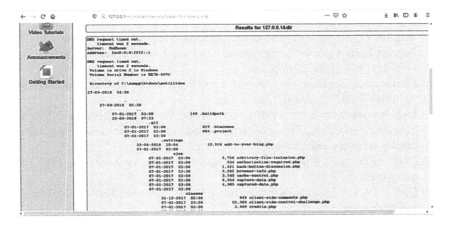

FIGURE 12.22 Output of OS command.

12.5 MITIGATION OF INJECTION VULNERABILITIES

To prevent an injection vulnerability in web applications, use the following mitigation steps:

a) Keep all data and information separate from commands and queries. For this purpose, disable the direct interaction of the attacker with the interpreter, or use a parameterized interface or migrate to object relational mapping tools.
b) Implement input validation at the server side. Properly manage all the input parameters.
c) Escape unnecessary special characters that are using the specific escape syntax for that interpreter.
d) Use SQL controls such as LIMIT and others within the SQL queries, to prevent the disclosure of information and records available in the database to prevent SQL injection.

12.6 BROKEN AUTHENTICATION AND SESSION MANAGEMENT

Attackers have easy access to millions of valid username and password combinations which are easily available on Internet for default admin-level accounts, credential stuffing, automatic brute force attacks and dictionary attack tools to exploit a broken authentication vulnerability. In session management-level attacks, a misuse of sessions, similar sessions, and unexpired session tokens are used by cybercriminals.

To identify the broken authentication vulnerability, there are lots of authentication weaknesses in the web application, including:

a) Permits automatic attacks such as credential stuffing, where the attacker has a long list of general valid usernames and passwords used by the users.
b) Permits brute force or other type of automated attacks, such as a dictionary-level attack, scripts, exploits, etc.
c) Permits weak, default, easy, or well-known passwords such as "admin/12345," "admin/admin," "test/test," "user/user."
d) Uses an ineffective or weak password-recovery mechanism and forgotten password procedure, like knowledge-based answers, one-time password, secondary recovery mail, mobile number, etc.
e) Uses clear text (unencrypted), weak encrypted or weak hashed slated passwords.
f) Don't have multifactor authentication or ineffective multifactor authentication.
g) Disclosure of session IDs in the URL.
h) Inefficient rotation of session IDs and tokens after successful authentication.
i) Does not properly validate or expiration of session IDs. User sessions or authentication tokens are improperly invalidated during the logout or a period of inactivity on the web application.

12.6.1 Exploitation of Broken Authentication Vulnerability

To exploit the broken authentication, we chose a vulnerable web application named "student zone," accessible on a local host, Figure 12.23

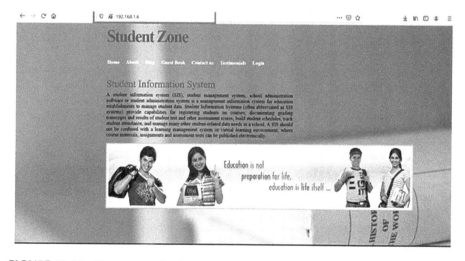

FIGURE 12.23 Home page of vulnerable web application accessible on local host.

Web Security Vulnerabilities 201

STEP 1: Going to ➔ home page (http://192.168.1.6),
STEP 2: Going to ➔ log-in page, Figure 12.24

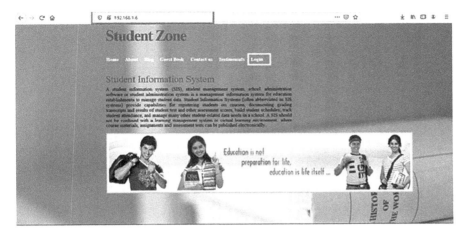

FIGURE 12.24   Home page of vulnerable web application.

STEP 3: At http://192.168.1.6/log-in/index.php page, enter any combination of user authentication credentials, like username: TEST and Password: TEST, Figure 12.25

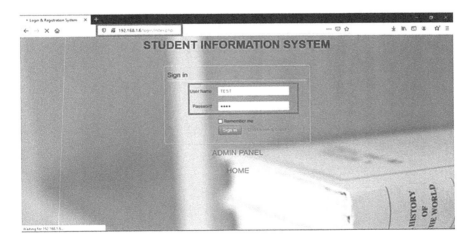

FIGURE 12.25   TEST/TEST as an input in the log-in field.

STEP 4: Using BURP Suite Professional v2.1.06 Web Security & Penetration Tool, which have the capability to intercept the http traffic (request and response). In Proxy → Intercept Tab (Raw http message), Figure 12.26:

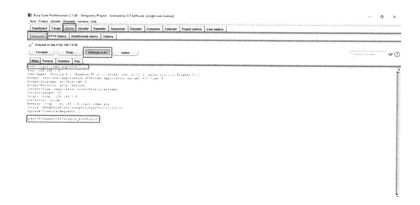

FIGURE 12.26 Intercept the http request/response.

STEP 5: Send these received details to Intruder tab, Figure 12.27.

FIGURE 12.27 Send the http response to intruder in the web application.

Web Security Vulnerabilities   203

STEP 6: Configure the payload on target 192.168.1.6 and port 80, Figure 12.28.

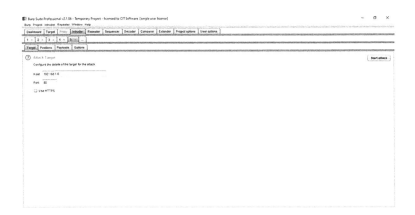

FIGURE 12.28   Configure the payload on port.

STEP 7: In Intruder Section → Payload Position → Select the Attack type (Cluster bomb), Figure 12.29.

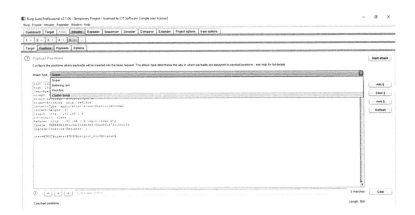

FIGURE 12.29   Intruder selection on payload.

STEP 8: In Intruder Section → Payload Position → Add payload markers at user and pass fields, Figure 12.30.

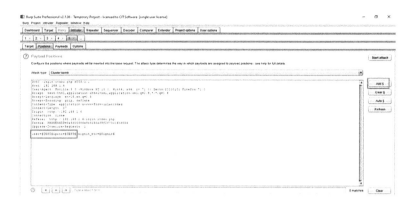

FIGURE 12.30 Intruder selection on payload with marked users.

STEP 9: In Intruder Section → Payloads→ Payload set → 1 and Payload type → Simple list of usernames, Figure 12.31.

FIGURE 12.31 Payload setup and simple type list generation.

Web Security Vulnerabilities

STEP 10: In Intruder Section → Payloads→ Payload set → 1 and Payload option (Simple List)→ Add from List → Usernames (List of default usernames), Figure 12.32.

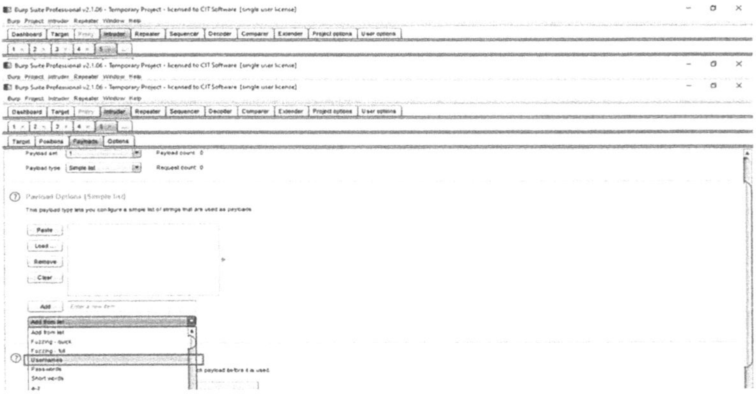

FIGURE 12.32   Payloads of huge list of usernames.

STEP 11: In Intruder Section → Payloads→ Payload set → 2 and Payload type → Simple list, Figure 12.33.

FIGURE 12.33   Select payload 2 as simple list.

206 Cybersecurity

STEP 12: In Intruder Section → Payloads→ Payload set → 2 and Add from List → Passwords (List of default passwords), Figure 12.34.

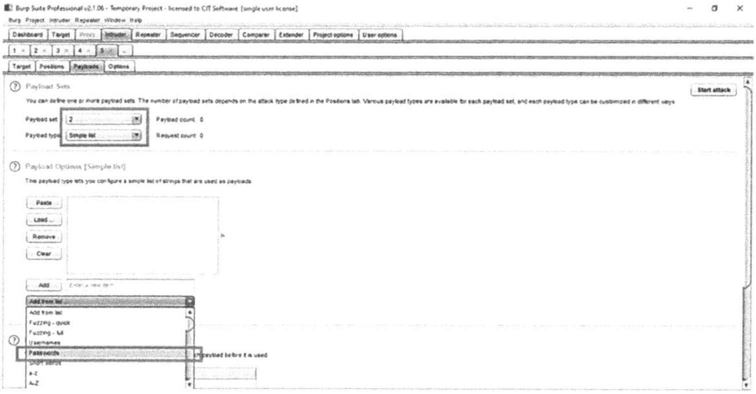

FIGURE 12.34 Payload 2 of huge password list.

STEP 13: In Intruder Section → Start Attack (Brute Force Attack), Figure 12.35.

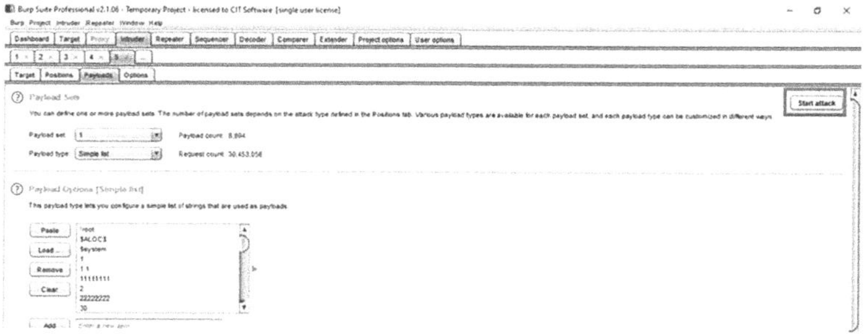

FIGURE 12.35 Start the brute force attack.

Web Security Vulnerabilities 207

STEP 14: In Attack Result Section, 10th row of request in which Payload1(Username):user and Payload2 (Password): user with status **302,** Figure 12.36.

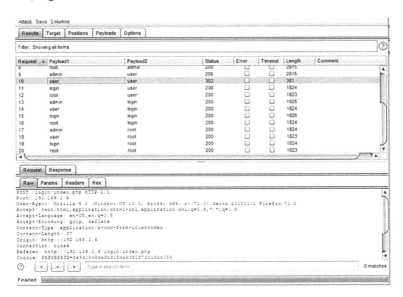

FIGURE 12.36 Username and password set matched at row 10.

STEP 16: HTTP Response message with 302 found, Figure 12.37.

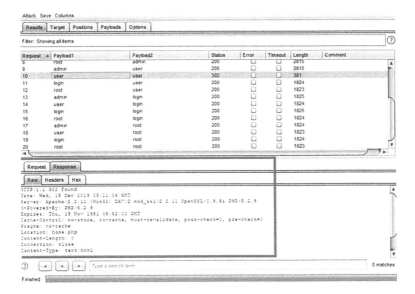

FIGURE 12.37 HTTP response 302 found.

STEP 16: Now in log-in page (**http://192.168.1.6/log-in/index.php**), enter Username: user and Password: user and Sign in, Figure 12.38.

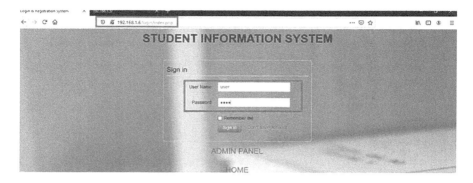

FIGURE 12.38 Log-in as found username: User and password: user.

STEP 17: Authentication Bypass Successful (**http://192.168.1.6/log-in/home.php**), Figure 12.39.

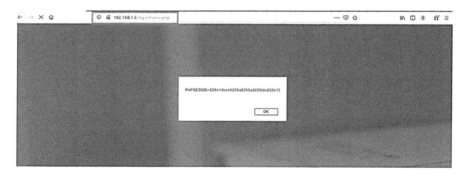

FIGURE 12.39 Successful log-in.

In this way an attacker can exploit the broken authentication vulnerability, if the user or developer chooses a weak credential, hard coded credential, or easily guessable passwords, using a manual guessable method or using tools such as OWASP ZAP or the Port Swigger Burp Suite.

12.6.2 MITIGATION OF BROKEN AUTHENTICATION AND SESSION MANAGEMENT

A web application developer can use following methods to mitigate the broken authentication and session management vulnerability:

a) Implement multifactor or two-way authentication to prevent automated attacks such as credential stuffing, brute force, dictionary attack, and stolen credential reuse attacks.

Web Security Vulnerabilities 209

b) When an application starts working as a live application, change the administrative-level password immediately. Do not use any default credentials, especially for administrative or super users.

c) Don't use hard coded credentials; implement a change password function for all users.

d) Use weak password check procedures, for example, testing new or changed passwords against a list of the top 10,000 worst, common, or previously used passwords.

e) Implement password-length and complexity features; also, choose a password policy for users.

f) Ensure registration and password recovery are optimally hard against account enumeration attacks.

g) Limit failed authentication attempts (a maximum of three log-in attempts), to reduce the possibility of an automated attack. Log all the failures and send the alert to the administrators as well as legitimate user when failed log-in attempts are identified or detected.

h) Use a secure, server side, built-in session manager that generates a new random session ID with higher entropy after a successful log-in.

i) Session IDs must not be clearly visible in the URL field. All session IDs should be invalidated after the log-out, or in an idle state, or after the absolute timeout.

12.7 SENSITIVE DATA EXPOSURE

Attackers are more interested in passive attacks, because these types of attacks hide their footprints. If information is transmitted from source to receiver while it is in transit, or from the user's client, e.g., a browser, an attacker can easily hack the information being transmitted, if it is in plain text or weakly encrypted.

12.7.1 IDENTIFICATION OF SENSITIVE DATA EXPOSURE VULNERABILITY

The following are steps to identify the vulnerability of sensitive data exposure:

a) Data transmission in clear plain text. This is related to protocols such as HTTP, SMTP, and FTP.

b) Use of weak or default cryptographic algorithms or any older vulnerable code.

c) Use of default cryptographic keys, weakly generated cryptographic keys, or missing rotations of cryptographic keys.

d) Enycryption is not used. For example, browser directories or encrypted headers are missing.

e) User agents or mail client do not verify the received server side certificates of integrity.

To mitigate this vulnerability, strong protection must be present for data in transit and data at rest. For example, sensitive personal data such as passwords, financial information such as credit/debit card numbers, health records of users, and business secrets require extra levels of protection, especially when the data and information

210   Cybersecurity

fall under various privacy laws or other legal mandates and international standardization, for example, the EU's General Data Protection Regulation (GDPR), financial data protection such as the PCI Data Security Standard (PCI DSS), information security management system (ISMS) like ISO/IEC 27001:2013, etc.

12.7.2 MITIGATION OF SENSITIVE DATA EXPOSURE VULNERABILITY

To mitigate or prevent sensitive data exposure vulnerability, the following are the minimal mitigation steps:

a) Classify the type of data or information as top secret, secret, confidential, sensitive, private, public, internal, etc.
b) Process, store, or transmit the data and information as per the classification. Identify which data or information is private or sensitive, according to the various privacy laws, legal mandates and regulatory requirements, and international standardization, or as per business needs. Apply security controls as per the data and information classification.
c) If it is not required to store sensitive data, than discard or delete it as soon as possible.
d) Encrypt all data and information in transit with secure protocols and other security parameters.
e) Use encryption for directives such as HTTP Strict Transport Security (HSTS).
f) Disable caching for responses that contain sensitive data and information.
g) Store passwords and other credentials using strong encryption and salted hashing functions.
h) Verify the effectiveness of configuration and settings regularly.

12.8 EXTERNAL ENTITIES OF XML (XXE)

Extensible Markup Language (XML) is a markup language that defines a set of rules for encoding documents in a format that it is readable both by humans or machines.

12.8.1 IDENTIFICATION OF XML EXTERNAL ENTITIES VULNERABILITY

A web application or XML-based web service may experience this vulnerability if:

a) The web application accepts XML documents directly, XML documents or a malicious shell from untrusted sources, or can insert untrusted data into XML documents, which is parsed by an XML processor.
b) Any of the XML processors in the web application has DTDs (document type definitions enabled.
c) The application uses Security Assertion Markup Language (SAML) for identity processing for single sign-on (SSO) purposes.
d) The Simple Object Access Protocol (SOAP) used in the application is prior to version 1.2. If XML entities are being passed to the SOAP framework, it may be susceptible to XXE attacks.

Web Security Vulnerabilities   211

e) If the application may vulnerable to XXE attacks, then the application may also be vulnerable to a billion laughs attack or DoS attacks. Here, a billion laughs attack is a type of DoS attack, aimed to the parsers of XML documents. It is also known as an XML bomb or as an exponential entity expansion (EEE) attack.

12.8.2 MITIGATION OF XML EXTERNAL ENTITIES

To mitigate the XML external entities vulnerability, developers should take following steps:

a) Upgrade or patch all the XML processors and libraries to their available latest version used in the web application or on the installed operating system.
b) Also use dependency checkers to update the SOAP to more available higher version.
c) Whenever possible, use less complex data formats, such as JSON.
d) Avoid the serialization of sensitive data and information.
e) Disable XML external entity and document type definitions (DTD) processing in all the XML parsers used in the web application.
f) Use positive whitelisting and server side input validation, and use filtering to prevent data within the headers, XML documents, or nodes.
g) Validate the incoming XML or any other file upload functionality using proper validation.
h) Static application security testing (SAST) tools can be used to detect an XXE vulnerability in source code. This as well as a manual code review is the best way to identify this vulnerability.
i) Also use API security gateways, virtual patching, or firewalls to monitor, detect, and block XXE attacks.

12.9 BROKEN ACCESS CONTROL

If a web application is vulnerable due to broken access controls, this means that a local user can work with more privileges like standard or admin users.

12.9.1 IDENTIFICATION OF BROKEN ACCESS CONTROL VULNERABILITY

Common access control vulnerabilities can be identified as:

a) Allowing the primary key or any id to be changed to another user's record, which is permitted to view or edit other user's account.
b) Bypassing access control mechanism by modifying the URL, state of internal application, HTML page, or attack through a custom API tool.
c) Elevation of privilege, acting as an admin user when logged in as a user or acting as a user without being logged in.
d) Force browsing as an unauthorized user to authenticated pages or to any privileged page as a standard user.

12.9.2 Prevent and Mitigate Broken Access Control

To prevent and mitigate the broken access control, take the following steps:

a) Domain models should be enforced for unique application business limit requirements.
b) Web roots must not present the web server directory listing and backup files.
c) In case of access control failures, alerts and logs must be maintained to inform the administrator.
d) Limit the application and control the access to minimize the harm from the tools of an automated attack.
e) Session tokens must be invalidated on the server after the log-out from an application or after a time limit.

12.10 MISCONFIGURATION OF SECURITY OPTIONS

Security misconfiguration is a common issue in web application security. It can happen at any level of the web application configuration, such as the web server, application server, platform, network services, frameworks, database, custom code, installed virtual machines, containers, or storage. A developer can check misconfiguration manually, but automated tools or web security scanners are more useful for detecting the security misconfigurations, use of default accounts or default configurations, unnecessary services or ports, etc.

Attackers will generally try to attempt and exploit the unpatched flaws or access default accounts, directories, unprotected files, unused pages, etc., to gain unauthorized access or to obtain the footprint of the system. These flaws generally give the attackers unauthorized access to data, information, or the functionality of the system. It will result in compromise of the system.

12.10.1 Identification of Security Misconfiguration Vulnerabilities

The application might be vulnerable if the web application has following issues:

a) Missing appropriate security hardening options across the whole web application or misconfigured permissions on cloud services.
b) Default accounts and their passwords are unchanged and enabled on run-time.
c) Improper error handling in the web application reveals lots of sensitive data or information.
d) The latest security features are disabled or not properly configured.
e) The security settings in the web application servers, frameworks, directories, libraries, databases, etc. are not set to secure values.
f) The server does not send security headers or they are not set to secure values.
g) The software, operating system, or application is vulnerable or out of date.
h) Without a repeatable or proper application security configuration process, computer resources are at a higher risk.

Web Security Vulnerabilities

213

12.10.2 Identification and Mitigation of Cross-Site Scripting (XSS)

The following are three forms of cross-site scripting (XSS) that generally target the browser of the user.

i. **Reflected Cross-Site Scripting (XSS):** The web application or API reflects invalidated user input as part of the HTML output. A successful XSS attack can allow the attacker to execute arbitrary HTML code and JavaScript in the victim's browser.

ii. **Stored Cross-Site Scripting (XSS):** The web application or API stores invalid unsanitized user input that is viewed or executed at a later time by an administrator or another user of web application. Stored XSS is considered as a high or critical risk in web applications.

iii. **DOM based Cross-Site Scripting (XSS):** JavaScript frameworks, single page web applications, and APIs that dynamically accept attacker controllable data to a page are known as a DOM-based XSS vulnerability. The application would not send attacker data to unsafe JavaScript APIs.

The XSS attacks used by the attackers for account takeover, session stealing, multi-factor authentication bypass, node replacement, or defacement of the web application, attacks against the browser of user system like key logging, malicious software downloads, and other client-side attacks.

12.10.3 Mitigation and Prevention of XSS Vulnerability

Preventing an XSS vulnerability requires the sanitization of untrusted data from the active browser content. This can be achieved by the following steps:

a) Escaping untrusted HTTP request data based in the HTML output such as body part, attribute, JavaScript, URL, or CSS can resolve reflected and stored XSS vulnerabilities.

b) Using software or web application frameworks that automatically escape cross-site scripting (XSS) by design, such as Ruby on Rails, appropriately handle the XSS use cases.

c) Applying context-sensitive encoding when modifying the document in browser on the client side acts against DOM XSS.

d) Enabling a content security policy (CSP) as a mitigating control against cross-site scripting (XSS).

12.10.4 Identification and Mitigation of Insecure Deserialization

Exploitation of deserialization is more difficult, compared to the detection of vulnerabilities; manual tampering with user data may give the existence of this vulnerability.

Some tools can discover deserialization flaws, but a manual method is needed to validate the problem or vulnerability.

214 Cybersecurity

The impact of deserialization flaws could pose serious hazards for any business application. An attacker can tamper with the price or quantity of products, or perform arithmetic operations on business pages.

12.11 IDENTIFICATION OF INSECURE DESERIALIZATION

Web applications and APIs may be vulnerable if they deserialize hostile or tampered-with objects provided by an attacker.

12.11.1 PREVENTION OF INSECURE DESERIALIZATION

To prevent this vulnerability, the web architecture or framework, serialized objects from untrusted sources are not accepted, or serialization mediums are used that allow primitive data types only. Also consider the following points:

a) Enforce strict data types and constraints during the deserialization before object creation, as the code expects a definable set of classes.
b) Use integrity checks, such as digital signatures, on any serialized objects to prevent hostile object creation or data tampering.
c) If possible, then isolate and run code that deserializes in low privilege environments.
d) Restrict or monitor incoming and outgoing data on networks that have been connected from containers or servers that perform deserialization.
e) Monitor the log of tampering and deserialization, alerting if a user tampers with the run-time data regularly and constantly.

12.11.2 IDENTIFICATION AND MITIGATION OF USING COMPONENTS WITH KNOWN VULNERABILITIES

The already written exploits are easy to find on the Internet for many known vulnerabilities. Some vulnerabilities require a more concentrated effort to develop a new custom exploit.

Some of the largest breaches to data and information have relied on exploiting known vulnerabilities in the components. If you are protecting the IT assets of an organization, always keep in mind that risks related to this vulnerability should be at the top of the list.

This vulnerability occurs when components such as framework and libraries used within the web application and mostly executed with full privileges and with known vulnerabilities. If a vulnerable component is exploited, it makes it very easy for a hacker to cause a serious data breach or server takeover.

12.11.3 IDENTIFICATION OF COMPONENTS WITH KNOWN VULNERABILITIES

A web application may have this vulnerability if:

a) A developer does not know the versions of all of the frameworks, components, tools, etc. used in the web application. Most of the products have different vulnerability notes or exploits. These exploits are easily available on the Internet.

Web Security Vulnerabilities   215

b) A computer resource is out of date, vulnerable, or unsupported. This includes the application server, operating system, database management system (DBMS), code, APIs, components, libraries, and run-time environments.

c) A developer does not scan the vulnerabilities regularly and does not have the knowledge or updates regarding latest security notes and bulletins related to the components you use in the application.

d) A developer does not fix or upgrade the components, frameworks, platform, and other dependencies regularly.

e) A developer team does not test the compatibility of updated, upgraded, or patched libraries.

12.11.4 MITIGATION OF COMPONENTS WITH KNOWN VULNERABILITIES

To mitigate and prevent these types of vulnerabilities, there should be a formal patch management process, and other steps may be adopted as follows:

a) Remove unused framework, components, unnecessary features, dependencies, functions, files, and other documentation.

b) Continuously monitor and make an inventory of the version's client-side and server-side components and their dependencies, using automated tools or a manual method.

c) Download the components from official and authentic sources only.

d) Downloading must be performed over the secure links.

e) Use an integrity check (checksum) of downloaded packages to reduce the chance of modification or the inclusion of malicious components.

f) Remove the libraries and components that are unmaintained or for which the service provider does not provide security patches for older versions.

g) Every organization and their developer team must ensure that there is a proper patch management plan for monitoring and applying updates, patches, and hotfixes or configuration changes for the lifetime of the application. Also, patch all the latest vulnerabilities regularly in a daily, weekly, monthly or quarterly manner that is suitable or as per the classification of the data (secret, confidential, internal, public, etc.).

h) The developer team must conduct an internal audit of the IT asset related to web applications. Also, conduct a third-party audit as per legal mandates or as and when required by qualified and empaneled cybersecurity auditors. It may be conducted in online and offline environments.

12.12 INSUFFICIENT LOGGING AND MONITORING VULNERABILITIES

The exploitation of insufficient logging and monitoring is the main reason for most of the major incidents. This vulnerability occurs when an attacker's activities on your web application are hidden and not traceable. Most successful attacks start with vulnerability identification or probing. Allowing such types of continued attack probes can raise the risk of successful exploitation on the web application.

216   Cybersecurity

12.12.1 IDENTIFICATION OF INSUFFICIENT LOGGING AND MONITORING VULNERABILITIES

A web application may have this vulnerability due to the following reasons:

a) Events and logs like successful log-ins, failed log-in attempts, and sudden high-value transactions, random input, and queries are not logged and not properly monitored.
b) Inadequate, unclear, default warnings and error-log messages.
c) Different logs of a web application are not properly monitored or reviewed for suspicious activity.
d) Local storage of logs.
e) Ineffective and inappropriate alert thresholds and response escalation processes.
f) Vulnerability assessment and penetration testing and security scans do not trigger alerts.

12.12.2 INSUFFICIENT LOGGING AND MONITORING VULNERABILITIES

To prevent this vulnerability, a developer should adopt the following preventive measures:

a) Ensure all events, such as log-in, failed log-in attempts, limit wrong log-in attempts, access control failures, or input validation failures should be logged to identify suspicious or malicious activities and stored for a sufficient length of time to allow for an investigation or delayed forensic analysis.
b) Establish or adopt an incident response, business continuity, or recovery plan. This is a mandatory document as per legal mandate and various international standards of cybersecurity.
c) Ensure that logs are generated in a format that should be easily understandable to an authority with proper centralized log management solutions.
d) Use commercial and open source application protection frameworks to monitor logs, such as OWASP AppSensor, web application firewalls, and log correlation application and software with custom dashboards for alerting and monitoring activities.

12.13 WEB SECURITY STANDARDS

Web applications must be reviewed and tested for security vulnerabilities against well-defined and internationally adopted web security standards. Applications that store, process, and transmit the data and information or provide access to sensitive data and information must be tested to an appropriate level and adopt baseline security features.

The following are web security standards and vulnerability references that must be known to web application developers.

i. **Open Web Application Security Project (OWASP) Top Ten:** The OWASP Top 10 is a web security standard document for the IT team, developers, and

Web Security Vulnerabilities

web security experts for web application security. It provides awareness of the most critical security risks to web applications. It is a good choice or first step towards changing the software development culture to secure software development within the organization, to produce more secure web applications.

ii. **SANS (SysAdmin, Audit, Network, Security) Top 25 Software Errors:** This provides the top 25 software errors that may result in critical risk to the web application. SANS Top 25 is a list of the most dangerous software errors. These are errors that can result in critical vulnerabilities that can allow attackers to steal data, modify information, or completely take over web applications.

iii. **The CWE (Common Weakness Enumeration) Top 25 Weaknesses:** This is a list of the most common and impactful web security vulnerabilities experienced by different organizations over the previous two years. These weaknesses are more dangerous because they are often easy to find and exploit and can allow attackers to completely take over a web application or software system. To reduce the vulnerabilities in the web application, the CWE Top 25 is a valuable community resource that can help web developers, security testers, users, web security auditors, project managers, researchers, and educators provide insight into critical web vulnerabilities and current security issues.

iv. **The Web Application Security Consortium (WASC):** This is a nonprofit international group of experts, professionals, industry practitioners, researchers, and organization representatives who develop open source and widely accepted best practices of security standards for the World Wide Web. WASC presents ideas to the web security community through various articles and releases technical information, security guidelines, and other useful documentation for computer researchers, organizations, educational institutes, governments, web application developers, IT security professionals, and software vendors all over the world, to provide assistance for the challenges in the field of computer security.

v. **Common Vulnerability Exposure (CVE):** This is a list of entries with an identification number for each vulnerability, a description of the vulnerability, and at least one public reference for the publicly known IT security vulnerabilities. CVE Entries are used for IT products, software, and IT services around the world.

vi. **OWASP Web Application Penetration Testing Checklist:** This is a penetration testing checklist developed by OWASP. It can be used by developers at the time of vulnerability assessment and penetration testing. The components of the penetration testing checklist promote consistency among both internal testing teams and external vendors. The checklist includes an RFP Template, Benchmarks, and Testing Checklist.

12.14 CONCLUSION

There is no one well-established platform or framework for web security. New vulnerabilities come into existence as a zero day vulnerability and enable attackers to attack the various web applications that have this type of vulnerability.

A web security team cannot go for maximum security. Cybersecurity is an ongoing process. You identify the vulnerability and patch it. Then again, and the process is repeated throughout the lifetime of web application.

A developer team may adopt the following methodology to enhance the cybersecurity posture of a web application and related frameworks:

a) Identify the security requirement of the application. Web Application Security Standards such as OWASP, as mentioned earlier, recommend application security practices or the mitigation of common vulnerabilities. This standard provides a baseline for testing web application vulnerabilities and also provides security recommendations for developers for secure development. Developers must also aware of other security references and standards, such as the SANS Top 25 errors, the CWE Top 25, the Web Application Security Consortium (WASC), etc.

b) Develop an application security architecture from the beginning of the web application project. Design security from the start, not after the release of the product.

c) Use a set of required security controls that simplifies the development of secure web applications.

d) Adopt a secure development life cycle.

e) Educate developers about web security through training, vulnerable web applications, hands-on experience, web security references, and awareness sessions.

f) Perform internal and external It security audits by certified auditors of computer resources used for web applications.

REFERENCES

1. SQL Injection, https://portswigger.net/web-security/sql-injection
2. Types of SQL Injection, https://www.acunetix.com/websitesecurity/sql-injection2
3. NOSQL Injection, https://www.netsparker.com/blog/web-security/what-is-nosql-injection
4. Expression Language (EL) or Object Graph Navigation Library (OGNL) injection, https://vulncat.fortify.com/en/detail?id=desc.dataflow.java.ognl_expression_injection
5. Test Vulnerable Web Application, http://testphp.vulnweb.com
6. OWASP Application Security Verification Standard, https://owasp.org/www-project-application-security-verification-standard
7. OWASP Cheat Sheet Series, https://owasp.org/www-project-cheat-sheets
8. Burp Suite Professional Tool, https://portswigger.net/burp
9. SANS Top 25 Software Errors, https://www.sans.org/top25-software-errors
10. The Common Weakness Enumeration Top 25, http://cwe.mitre.org/top25/archive/2020/2020_cwe_top25.html
11. Web Application Security Consortium, http://www.webappsec.org
12. Common Vulnerability Exposure, https://cve.mitre.org
13. A Vulnerable Web Application of Acunetix Tool, http://testphp.vulnweb.com
14. Web Application Security Standards, https://cuit.columbia.edu/sites/default/files/content/Web%20Application%20Security%20Standards%20and%20Practices.pdf
15. OWASP Web Application Penetration Testing Checklist, https://owasp.org/www-project-web-security-testing-guide/assets/archive/OWASP_Web_Application_Penetration_Checklist_v1_1.pdf

INDEX

A

access control, 11, 31, 59, 81, 143, 151, 156, 212
accountability, 78, 80
anomaly detection, 21, 170
anonymity, 33, 38, 78
API, 56, 60, 213
application security, 34, 217
Arduino Uno, 124, 125
ARM, 4
artificial intelligence (AI), 50, 73, 90
audit-based access control, 153
authenticity, 5, 136
authorization, 5, 11, 137, 152
automation, 24, 33, 84
availability, 31, 55, 68, 78, 136

B

bacterial blight, 120
bandwidth, 71
base station, 135, 140
behavior-based access control, 153
Bell-LaPadula model, 6, 7
Biba model, 3, 8
blind SQL injection, 186
blind time-based SQL injection, 187
BPMN, 3
broken access control, 185, 211, 212
brown spot, 120, 121
Browser history, 100

C

CBM, 4
CCHIT, 157
CDI, 9, 10
CHAOS, 143, 169
Chinese Wall security, 11
CIA, 31
Clarke-Wilson model, 9, 10
cloud computing, 44, 47, 53, 59
CLPA, 144
code tampering, 11, 13
common vulnerability exposure, 217
compelled data items, 10
confidentiality, 4, 7, 31, 78, 135, 142
COVID-19, 44

cross-site scripting, 36, 213
crowd response, 99
cryptography, 4–6, 11, 12, 52, 83, 139, 144, 155
CVE, 217
cybercrime, 30, 39, 89, 101
cybersecurity, 4, 18, 20, 31, 33, 36, 39, 218
cyberspace, 36, 38

D

data availability, 136
database forensics, 86
data integrity, 5, 136
data privacy, 78, 137
DBSy, 3
DDoS, 22, 23, 35, 48, 59, 105
deep neural networks, 172
denial of service (DoS), 22, 23, 35, 48, 53, 104, 105, 139, 141, 164, 184, 185, 211
digital copyright, 30
digital forensics, 84–86, 90, 97, 98, 100
discretionary access control, 151
disk forensics, 86
document type definitions, 210, 211
domain name system, 49
Dumpzilla, 100

E

e-health, 105, 133, 135, 139
electronic health record, 150, 152, 157
email forensics, 86
encrypted disk detector, 99
error-based SQL injection, 186, 188, 189
Ethernet Shield, 125, 128
ethics, 27, 36, 40
ExifTool, 99

F

FAW, 99
flood attacks, 138
flooding, 22, 23, 139
FNR, 165, 166
forensic investigator, 95, 99
ForensicUserInfo, 100
framework, 2, 6, 10, 47, 60, 74, 98, 143, 144, 153, 169, 210

220 Index

G

GPS, 29, 30

H

HashMyFiles, 99
head-mounted displays, 14
heartbeat, 134, 139
heartbleed scanner, 99
HEBM, 143
HER, 150
HIPAA, 157
HRU security model, 10
Hyper-Text Transfer Protocol, 49

I

IaaS, 55
ICT, 67, 68, 77
identity theft, 11, 12, 39, 101, 104, 185
IDS, 18, 20–22
implant node, 133
industrial Internet of Things, 48, 50
Industry 4.0, 50, 84
information security, 31, 34, 149, 210
information technology, 30, 45, 54, 132, 157
Information Theft, 12, 13
injection, 185, 186
injection vulnerabilities, 187, 199
integrity, 5, 18, 31, 78, 101, 136
integrity verification procedure, 10
Internet of everything, 84, 104
Internet-of-Things (IoT), 44, 48, 49, 51, 68, 70, 71, 84, 104, 123, 132
Internet of Vehicles (IoV), 77
IoT BAI, 104

J

jamming, 139

K

Kali Linux, 100
KDD99, 164
key distribution, 137, 140–142
key management, 6, 141, 144

L

LEA, 143
Leaf smut, 120
lightweight directory access, 185
location-based services, 72
location-restricted mobility, 72
long-term evaluation (LTE), 72

LRR, 72
LSTM-RNN, 172

M

Magnet RAM Capture, 99
malware, 11, 32, 114
malware forensics, 86
mandatory access control, 151
man-in-the-middle, 139, 143
manual attacks, 24
MCS, 68
memory forensics, 87
mobile crowd sensing, 68
mobile phone forensics, 87
mobility, 66
Mobility as a Service (MaaS), 77
moisture sensor, 125

N

NDAE, 19
network forensics, 86
network miner, 99
network security, 31, 33, 34, 59, 161
neural network, 163, 171, 172
NFI Defraser, 99
NIDS, 19
NMAP, 99
non-repudiation
NoSQL injection, 187
NVD, 47

O

OASIS role-based access, 154
operational security, 34
OS command injection, 187
OWASP, 11, 51, 190, 216

P

PaaS, 55
Paladin, 100
particle swarm optimization, 162
path-restricted mobility, 72
peacefulness rule, 8
pedestrian-to-infrastructure, 73
penetration testing, 217
pH, 123, 126
phishing, 13, 33, 39, 85, 185
PIF, 4
P-RBAC, 153
privacy, 29, 36, 46, 77–80, 137, 150
    by design, 80
    enhancing techniques, 79
    of location, 79

Index

221

PRM, 72
probing attacks, 23
PSO, 162, 163

R

R2L, 164
RAM capturer, 99
random forest, 170
RBAC, 151
replay attack, 138
reverse engineering, 14, 101
role-based access control, 151
rule-based access control, 151

S

SaaS, 55
SAML, 210
SANS, 97
SCADA, 3
Scareware, 32
SD-IoT, 48
security model(ing), 2, 6
security services, 4
security strategy, 2
selective forwarding, 138
sensitive data exposure, 209
Shellshock Scanner, 99
signature-base detection, 21
Sinkhole Attacks, 138
Sleuth Kit, 100
smart mobility, 66, 70, 75
SOAP, 210
solid star property rule, 8
SQL, 185
Star Confidentiality Principle, 7
Star Integrity Principle, 9
stateful protocol analysis, 22
static application security testing, 211
strong star confidentiality principle, 7
SVM, 167
Sybil Attacks, 138

T

tampering, 139
TCS3200 Color Sensor, 127
technoethics, 28

telemedicine, 139
threat, 2
Toolsley, 100
TPR, 165
transformation procedures, 9
Trojans, 23
trust management, 29

U

UDI, 9
UMM, 72
unconstrained information things, 9, 10
uncontrollable mobility, 72
union-based SQL injection, 186
unrestricted mobility, 72
URM, 72
USB Write Blocker, 99
U2R, 164

V

vehicle-to-infrastructure, 73
vehicle-to-pedestrian, 73
vehicle-to-vehicle, 73
VIP, 73
virtual reality, 14
viruses, 23
volatility, 100
vulnerabilities, 45

W

WBAN, 143
Wi-Fi, 73
WindowSCOPE, 99
wireless forensics, 86
wireshark, 99
wormholes, 138
worms, 23
WPAN, 143
WSN, 105, 144

X

XACML, 152, 154
XML, 185, 210

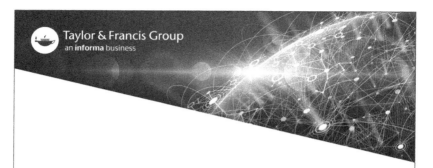